Station Activities

for Mathematics

Grade 6

PSMS

WALCH EDUCATION

Certified Chain of Custody
Promoting Sustainable
Forest Management
www.sfiprogram.org

SGS-SFI/COC-US09/5501

1   2   3   4   5   6   7   8   9   10

ISBN 978-0-8251-6450-7

Copyright © 2008

J. Weston Walch, Publisher

Portland, ME 04103

www.walch.com

Printed in the United States of America

Table of Contents

Introduction .... *v*

Materials List .... *viii*

Number and Operations 1

Set 1: Factors, Multiples, and Prime Factorization 1

Set 2: Fractions, Decimals, and Percents 8

Set 3: Adding and Subtracting Fractions 15

Set 4: Multiplying and Dividing Fractions 22

Set 5: Problem Solving with Fractions, Decimals, and Percents 29

Geometry and Measurement 37

Set 1: Appropriate Units of Measurement 37

Set 2: Volume and Surface Area 44

Set 3: Ratio, Proportion, and Scale 53

Set 4: Visualizing Solid Figures 60

Set 5: Problem Solving with Volume, Surface Area, and Scale 68

Algebra 75

Set 1: Patterns and Relationships 75

Set 2: Graphing Relationships 82

Set 3: Evaluating Expressions 91

Set 4: Solving Equations 98

Set 5: Proportional Relationships 105

Data Analysis and Probability 113

Set 1: Collecting, Organizing, and Analyzing Data 113

Set 2: Constructing Frequency Distributions 120

Set 3: Using Tables and Graphs 127

Set 4: Theoretical Probability 135

Set 5: Experimental Probability 142

Introduction

This book includes a collection of station-based activities to provide students with opportunities to practice and apply the mathematical skills and concepts they are learning. It contains five sets of activities for each of the four strands: Number and Operations; Geometry and Measurement; Algebra; and Data Analysis and Probability. You may use these activities in addition to the direct instruction lessons, or, especially if the pre-test or other formative assessment suggests it, instead of direct instruction in areas where students have the basic concepts but need practice. The debriefing discussions after each set of activities provide an important opportunity to help students reflect on their experiences and synthesize their thinking. It also provides an additional opportunity for ongoing, informal assessment to inform instructional planning.

Implementation Guide

The following guidelines will help you prepare for and use the activity sets in this book.

Setting Up the Stations

Each activity set consists of four or more stations. Set up each station at a desk, or at several desks pushed together, with enough chairs for a small group of students. Place a card with the number of the station on the desk. Each station should also contain the materials specified in the teacher's notes, and a stack of Student Activity Sheets (one copy per student). Place the required materials (as listed) at each station.

When a group of students arrives at a station, each student should take one of the activity sheets to record the group's work. Although students should work together to develop one set of answers for the entire group, each student should record the answers on his or her own activity sheet. This helps keep students engaged in the activity and gives each student a record of the activity for future reference.

Forming Groups of Students

All activity sets consist of four stations. You might divide the class into four groups by having students count off from 1 to 4. If you have a large class and want to have students working in small groups, you might set up two identical sets of stations, labeled A and B. In this way, the class can be divided into eight groups, with each group of students rotating through the "A" stations or "B" stations.

Assigning Roles to Students

Students often work most productively in groups when each student has an assigned role. You may want to assign roles to students when they are assigned to groups and change the roles occasionally. Some possible roles are as follows:

- **Reader**—reads the steps of the activity aloud
- **Facilitator**—makes sure that each student in the group has a chance to speak and pose questions; also makes sure that each student agrees on each answer before it is written down
- **Materials Manager**—handles the materials at the station and makes sure the materials are put back in place at the end of the activity
- **Timekeeper**—tracks the group's progress to ensure that the activity is completed in the allotted time
- **Spokesperson**—speaks for the group during the debriefing session after the activities

Timing the Activities

The activities in this book are designed to take approximately 15 minutes per station. Therefore, you might plan on having groups change stations every 15 minutes, with a two-minute interval for moving from one station to the next. It is helpful to give students a "5-minute warning" before it is time to change stations.

Since the activity sets consist of four stations, the above timeframe means that it will take about an hour and 10 minutes for groups to work through all stations. If this is followed by a 20-minute class discussion as described below, an entire activity set can be completed in about 90 minutes.

Guidelines for Students

Before starting the first activity set, you may want to review the following "ground rules" with students. You might also post the rules in the classroom.

- All students in a group should agree on each answer before it is written down. If there is a disagreement within the group, discuss it with one another.
- You can ask your teacher a question only if everyone in the group has the same question.
- If you finish early, work together to write problems of your own that are similar to the ones on the Student Activity Sheet.
- Leave the station exactly as you found it. All materials should be in the same place and in the same condition as when you arrived.

Debriefing the Activities

After each group has rotated through every station, bring students together for a brief class discussion. At this time you might have the groups' spokespersons pose any questions they had about the activities. Before responding, ask if students in other groups encountered the same difficulty or if they have a response to the question. The class discussion is also a good time to reinforce the essential ideas of the activities. The questions that are provided in the teacher's notes for each activity set can serve as a guide to initiating this type of discussion.

You may want to collect the Student Activity Sheets before beginning the class discussion. However, it can be beneficial to collect the sheets afterward so that students can refer to them during the discussion. This also gives students a chance to revisit and refine their work based on the debriefing session.

Materials List

Class Sets

- calculators
- rulers
- protractors
- scissors

Station Sets

- fraction circles
- algebra tiles and equation mats
- counters (50–100 chips, counters, beans, pennies)
- tiles (+/– 25 of each of several colors)
- integer chips
- rectangular prism
- regular shapes (triangle, square, pentagon, hexagon, heptagon, octagon, nonagon, decagon, dodecagon)
- variety of round objects
- string
- box/container 2 in. × 4 in. × 8 in.
- 64 1-inch cubes
- cylinder
- geoboards and rubber bands
- spinners
- Unifix or other connecting cubes
- colored cubes
- bags (fabric or opaque paper)

Ongoing Use

- index cards (need to be prepared according to specifications in teacher notes for many of the station activities)
- graph paper
- pencils
- highlighters of several colors
- pencils/markers of several colors
- pennies
- number cubes

Other

- bags of fun-size M&Ms®
- boxes of toothpicks

Overhead Manipulatives (optional)

- clock
- algebra tiles
- geometric shapes
- protractor
- tiles
- fraction circles
- spinners

Number and Operations

Set 1: Factors, Multiples, and Prime Factorization

Goal: To provide opportunities for students to develop concepts and skills related to factors, multiples, and prime factorizations

NCTM Standards, Grades 6–8

Number and Operations

Understand numbers, ways of representing numbers, relationships among numbers, and number systems: use factors, multiples, prime factorization, and relatively prime numbers to solve problems.

Student Activities Overview and Answer Key

Station 1

Students use a number cube to generate two-digit numbers. Then they work together to decide if each of the numbers is prime or composite. If the number is composite, they find the prime factorization. Finally, students describe the strategies they used to identify the numbers as prime or composite.

Answers: Answers will depend on the numbers that are rolled.

Possible strategies: Any two-digit even number is composite; any number ending in 5 is divisible by 5, so it is composite; recognize familiar primes, such as 11 and 13.

Station 2

Students are given 12 red tiles and 18 blue tiles. They are asked to arrange the tiles in rows so that each row contains the same number of tiles and so that each row contains only red tiles or only blue tiles. The rows must also be as long as possible. When students have arranged the tiles following these rules, they reflect on how their arrangement is related to the greatest common factor of 12 and 18.

Answers: Each row has 6 tiles. There are 2 rows with 6 red tiles in each row and 3 rows with 6 blue tiles in each row. The number of tiles in each row (6) is the greatest common factor of 12 and 18.

Station 3

Students use a highlighter to highlight all the numbers less than or equal to 100 that are multiples of 8. Then they use a different color to highlight all the multiples of 12. Students work together to check that the multiples are highlighted correctly. Then they look for common multiples (numbers highlighted in both colors) and identify the least common multiple.

Answers: Multiples of 8: 8, 16, 24, 32, 40, 48, 56, 64, 72, 80, 88, 96

Multiples of 12: 12, 24, 36, 48, 60, 72, 84, 96

Numbers highlighted in both colors (common multiples): 24, 48, 72, 96

The least common multiple of 8 and 12 is the smallest of the common multiples, 24.

Station 4

Students are given a set of cards with numbers on them. They choose two cards at random and work together to find the greatest common factor and least common multiple of the two numbers chosen. Students replace the cards, shuffle them, and repeat the process two more times. Then they reflect on the strategies they used.

Answers: Answers will depend upon the cards that are chosen.

Possible strategies: To find the greatest common factor, list all the factors of each number and choose the greatest factor that appears in both lists. To find the least common multiple, list several multiples of each number and choose the smallest multiple that appears in both lists.

Materials List/Set Up

Station 1	number cube (numbered 1–6)
Station 2	12 small red tiles; 18 small blue tiles
Station 3	2 highlighters (different colors)
Station 4	8 index cards with the following numbers written on them:
	4, 6, 8, 10, 12, 15, 16, 18

Discussion Guide

To support students in reflecting on the activities and to gather some formative information about student learning, use the following prompts to facilitate a class discussion to "debrief" the station activities.

Prompts/Questions

1. What are some ways to decide if a number is prime or composite?

2. Is it ever possible for an even number to be prime? Why or why not?

3. Suppose you had 98 red tiles and 66 blue tiles, and you wanted to arrange all of them in rows so that each row contained only red tiles or only blue tiles, and so that the rows were as long as possible. How could you find the number of tiles in each row without actually arranging all the tiles?

4. Can the least common multiple of two numbers ever be one of the given numbers? Explain.

Think, Pair, Share

Have students jot down their own responses to questions, then discuss with a partner (who was not in their station group), and then discuss as a whole class.

Suggested Appropriate Responses

1. Possible methods: Aside from 2, any even number is composite. A prime number cannot have a units digit of 5. Check to see if the number is divisible by any of the prime numbers that are smaller than it.

2. Yes; the number 2 is prime. Any other even number is not prime because it is divisible by 2.

3. Find the greatest common factor of 98 and 66.

4. Yes. For example, the least common multiple of 2 and 10 is 10.

Possible Misunderstandings/Mistakes

- Assuming that odd numbers are prime (e.g., stating that 39 is prime)
- Multiplying two numbers to find the least common multiple (e.g., stating that the least common multiple of 6 and 8 is 48)
- Forgetting that the greatest common factor of two numbers may be one of the numbers

Number and Operations
Set 1: Factors, Multiples, and Prime Factorization

Station 1

You will find a number cube at this station. You will use the number cube to create some two-digit numbers.

Roll the number cube two times. Write the numbers in the boxes below.

Work with other students to decide if this two-digit number is prime or composite. Write the answer on the line. _____

If the number is composite, write the prime factorization of the number on the line below.

_____

Repeat the process 2 more times.

Prime or composite? _____

If composite, prime factorization: _____

Prime or composite? _____

If composite, prime factorization: _____

Write at least three strategies you used to help you decide whether each number was prime or composite.

_____

_____

_____

_____

Number and Operations
Set 1: Factors, Multiples, and Prime Factorization

Station 2

At this station, you will find 12 red tiles and 18 blue tiles.

Work with other students to arrange the tiles in rows. You must follow these rules:

- Use all the tiles.
- A row can contain only blue tiles or only red tiles.
- Each row must contain the same number of tiles.
- The rows should be as long as possible.

Sketch your arrangement of tiles in the space below.

Work together to check that your arrangement matches all the rules.

How many tiles are in each row? _____

Explain how the number of tiles in each row is related to the greatest common factor.

_____

_____

_____

_____

Number and Operations
Set 1: Factors, Multiples, and Prime Factorization

Station 3

You will find two highlighters at this station. Use one highlighter to highlight all the numbers in the grid that are multiples of 8. Use the other highlighter to highlight all the numbers in the grid that are multiples of 12.

1	2	3	4	5	6	7	8	9	10
11	12	13	14	15	16	17	18	19	20
21	22	23	24	25	26	27	28	29	30
31	32	33	34	35	36	37	38	39	40
41	42	43	44	45	46	47	48	49	50
51	52	53	54	55	56	57	58	59	60
61	62	63	64	65	66	67	68	69	70
71	72	73	74	75	76	77	78	79	80
81	82	83	84	85	86	87	88	89	90
91	92	93	94	95	96	97	98	99	100

Work together to check that you have highlighted the multiples correctly.

Which numbers are highlighted in both colors? _____

What can you say about these numbers? _____

_____

Explain how to use your work to find the least common multiple of 8 and 12.

_____

_____

_____

Number and Operations
Set 1: Factors, Multiples, and Prime Factorization

Station 4

You will be given a set of index cards. Shuffle the cards and place them face down.

Choose two cards without looking. Turn the cards over. Write the two numbers below.

_____ _____

Work with other students to find the greatest common factor and least common multiple of the numbers.

Greatest common factor: _____

Least common multiple: _____

Put the cards back. Shuffle the cards. Then repeat the above process two more times.

Two numbers: _____ _____

Greatest common factor: _____

Least common multiple: _____

Two numbers: _____ _____

Greatest common factor: _____

Least common multiple: _____

Explain the strategies you used to find the greatest common factors.

_____

_____

_____

Explain the strategies you used to find the least common multiples.

_____

_____

_____

Number and Operations

Set 2: Fractions, Decimals, and Percents

Goal: To provide opportunities for students to develop concepts and skills related to fractions, decimals, and percents

NCTM Standards, Grades 6–8

Number and Operations

Understand numbers, ways of representing numbers, relationships among numbers, and number systems: work flexibly with fractions, decimals, and percents to solve problems; compare and order fractions, decimals, and percents efficiently and find their approximate locations on a number line.

Student Activities Overview and Answer Key

Station 1

Students use a number cube to create fractions. They work together to write each fraction as a decimal and as a percent.

Answers: Answers will vary depending upon the numbers rolled.

Possible strategies: To write the fraction as a decimal, divide the numerator by the denominator. To change the decimal to a percent, move the decimal point two places to the right (adding a zero if necessary), then add a percent sign.

Station 2

Students model decimals by shading squares of 10-by-10 grids. Students work as a group to ensure that each decimal is modeled correctly. Then they use the models to decide what fraction of each grid is shaded. In this way, students write equivalent fractions for each decimal.

Answers: 1. $^1/_2$; 2. $^7/_{10}$; 3. $^3/_4$; 4. $^1/_{20}$; 5. $^1/_{100}$; 6. $^{27}/_{100}$

Possible explanation: The fraction is the number of shaded squares (numerator) over 100 (denominator). After writing this fraction, simplify by dividing common factors out of the numerator and denominator.

Instruction

Station 3

Students are given eight numbers in the form of fractions, decimals, and percents. The numbers are written on index cards and students work together to decide how the numbers should be plotted on a number line. Then they determine which of the given numbers are equal.

Answers: $^1/_5 = 20\%$; $0.6 = ^3/_5$; $^3/_4 = 75\%$

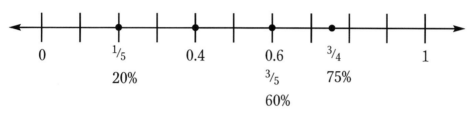

0	$^1/_5$	0.4	0.6	$^3/_4$	1
	20%		$^3/_5$	75%	
			60%		

Possible strategies: Each interval on the number line represents $^1/_{10}$ (or 0.1) and this may be used to help plot the decimals; convert the fractions and percents to decimals; compare the given numbers to familiar benchmarks.

Station 4

Students are given ten numbers in the form of fractions, decimals, and percents. The numbers are written on index cards and students work together to find pairs of cards that represent equal numbers. Then students write equality statements based on the pairs of matching cards.

Answers: $6\% = 0.06$; $25\% = ^1/_4$; $0.4 = ^2/_5$; $^1/_2 = 50\%$; $0.6 = ^3/_5$

Possible strategies: Start by finding familiar benchmarks (e.g., $^1/_2 = 50\%$); convert the numbers to decimals; convert to fractions with like denominators.

Materials LIst/Set Up

Station 1   number cube (numbered 1–6)

Station 2   none

Station 3   8 index cards with the following numbers written on them:

$^1/_5$, 20%, 0.4, 0.6, $^3/_5$, 60%, $^3/_4$, 75%

Station 4   10 index cards with the following numbers written on them:

6%, 0.06, 25%, $^1/_4$, 0.4, $^2/_5$, $^1/_2$, 50%, 0.6, $^3/_5$

Discussion Guide

To support students in reflecting on the activities and to gather formative information about student learning, use the following prompts to facilitate a class discussion to "debrief" the station activities.

Prompts/Questions

1. What are some different tools, drawings, or methods that you can use to help you work back and forth among fractions, decimals, and percents?

2. How do you convert a fraction to a decimal?

3. How do you convert a decimal to a percent?

4. Is it possible to write every fraction as a decimal? Why or why not?

Think, Pair, Share

Have students jot down their own responses to questions, discuss their responses with a partner (who was not in their station group), and then discuss as a whole class.

Suggested Appropriate Responses

1. 10-by-10 grids, number lines, etc.

2. Divide the numerator of the fraction by the denominator of the fraction.

3. Move the decimal point two places to the right (adding a zero if necessary) and add a percent symbol.

4. Yes. No matter what numbers appear in the fraction as the numerator and denominator, it is always possible to divide the numerator by the denominator (assuming the denominator is not zero), and this will convert the fraction to a decimal.

Possible Misunderstandings/Mistakes

- Misunderstanding the role of zero as a placeholder in decimals such as 0.05 (e.g., confusing this with 0.5)

- Incorrectly converting decimals to percents (e.g., writing 0.7 as 7%)

- Incorrectly converting fractions to decimals (e.g., $^2/_5$ = .25)

Number and Operations
Set 2: Fractions, Decimals, and Percents

Station 1

You will find a number cube at this station. Use the number cube to create fractions.

Roll the number cube two times. Write the two numbers in the boxes below to create a fraction.

Work with other students to write the fraction as a decimal and as a percent. Write the decimal and percent on the line below.

_____

Repeat the process four more times.

Explain how you wrote the fractions as decimals and percents.

_____

_____

_____

_____

Number and Operations
Set 2: Fractions, Decimals, and Percents

Station 2

Each 10-by-10 grid represents 1. Each column of the grid represents one-tenth. Each small square represents one one-hundredth.

Show each decimal by shading squares in the grid. Work as a group to make sure each decimal is shown correctly. Then decide what fraction of the grid is shaded. Write the fraction in the space provided.

1. 0.5 _____

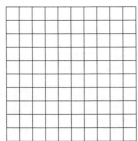

4. 0.05 _____

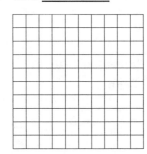

2. 0.7 _____

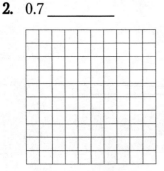

5. 0.01 _____

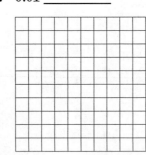

3. 0.75 _____

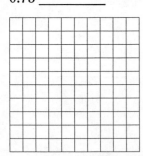

6. 0.27 _____

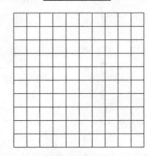

Explain how you decided what fraction was shown in each of the grids.

_____

_____

_____

Number and Operations
Set 2: Fractions, Decimals, and Percents

Station 3

You will be given a set of cards with the following numbers written on them.

$$\frac{1}{5} \qquad 0.4 \qquad \frac{3}{4} \qquad \frac{3}{5} \qquad 60\% \qquad 75\% \qquad 20\% \qquad 0.6$$

Work with other students to plot and label each number on the number line. Work as a group to make sure the numbers are all plotted correctly.

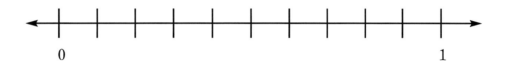

Which of the given numbers are equal? (*Hint:* Look for numbers that are plotted at the same point on the number line.)

_____

_____

Write at least three strategies you used to help you plot the numbers on the number line.

_____

_____

_____

_____

_____

_____

_____

_____

_____

Number and Operations
Set 2: Fractions, Decimals, and Percents

Station 4

You will be given a set of cards with the following numbers written on them.

$$\frac{3}{5} \qquad 0.4 \qquad 25\% \qquad 6\% \qquad \frac{1}{4} \qquad \frac{2}{5} \qquad 0.6 \qquad \frac{1}{2} \qquad 0.06 \qquad 50\%$$

Work with other students to find pairs of cards that show the same number. When you have paired up the cards, work as a group to check that the numbers in each pair are equal.

Write five statements that use an equal sign (=) to list the pairs of equal numbers.

_____

_____

_____

_____

_____

Write at least three strategies you used to help you decide which numbers were equal.

_____

_____

_____

_____

_____

_____

_____

Number and Operations

Set 3: Adding and Subtracting Fractions

Goal: To provide opportunities for students to develop concepts and skills related to addition and subtraction

NCTM Standards, Grades 6–8

Number Sense, Concepts, and Operations

Understand meanings of operations and how they relate to one another: understand the meaning and effects of arithmetic operations with fractions, decimals, and integers.

Student Activities Overview and Answer Key

Station 1

Students model the sum of two fractions using fraction circles. They work as a group to ensure that the sum is modeled correctly. Then they use the model to find the sum of the fractions.

Answers: 1. $\frac{3}{5}$; 2. $\frac{4}{10}$ or $\frac{2}{5}$; 3. $\frac{1}{2}$; 4. $\frac{5}{6}$; 5. $\frac{7}{8}$; 6. 1

Possible strategies: If the two fractions have the same denominator, count the total number of pieces used. If the two fractions have unlike denominators, place smaller fraction pieces on top of the model to show the same portion of a complete circle. Count the number of smaller fraction pieces used.

Station 2

Students use rectangular strips of paper to help them subtract fractions. They first fold the strips into sixteen equal sections. Given a subtraction problem, students work together to model the first fraction by coloring sections of the strip yellow. Within the yellow section of the strip, students color sections blue to model the fraction being subtracted. The remaining yellow sections model the answer to the subtraction problem.

Answers: The remaining yellow section of the strip shows the fraction $\frac{3}{8}$.

1. $\frac{7}{16}$; 2. $\frac{1}{4}$; 3. $\frac{5}{8}$; 4. $\frac{3}{16}$; 5. $\frac{7}{16}$

Station 3

Students are given two sets of cards with fractions written on them. They choose one card from each set and work together to find the difference of the fractions. At the end of the activity, students discuss the strategies they used to subtract the fractions.

Answers: Answers will depend upon the cards that are chosen.

Possible strategies: If a card has a mixed number, first convert it to an improper fraction. If the denominators of the two fractions are the same, subtract the numerators and keep the same denominator. If the denominators are different, find equivalent fractions with common denominators.

Station 4

Students use a number cube to generate pairs of fractions. They work together to add the two fractions in each pair. All students in the group should agree on each answer.

Answers: Answers will depend upon the numbers that are rolled.

Possible strategies: If the two fractions have the same denominator, keep the denominator and add the numerators to get the numerator of the sum. If the two fractions have different denominators, write each fraction as an equivalent fraction so that the equivalent fractions have common denominators. Then add the equivalent fractions.

Materials List/Set Up

Station 1   set of fraction circles

Station 2   yellow and blue highlighters
rectangular strips of paper, about one inch wide and one foot long (These may be cut from sheets of paper, or you can use strips of adding-machine tape.)

Station 3   10 index cards, prepared as follows:

- set of 5 cards, labeled "A" on the back with the following fractions written on the front: $\frac{3}{4}$, $\frac{7}{8}$, $1\frac{1}{3}$, $1\frac{1}{2}$, $2\frac{2}{3}$

- set of 5 cards, labeled "B" on the back with the following fractions written on the front: $\frac{1}{6}$, $\frac{1}{3}$, $\frac{3}{8}$, $\frac{1}{2}$, $\frac{5}{8}$

- Place the cards face-down at the station, so that only the backs (A or B) are visible.

Station 4   number cube (numbers 1–6)

Discussion Guide

To support students in reflecting on the activities and to gather some formative information about student learning, use the following prompts to facilitate a class discussion to "debrief" the station activities.

Prompts/Questions

1. What are some different tools, drawings, or methods that you can use to help add and subtract fractions?

2. How do you add two fractions with like denominators?

3. How do you add two fractions with unlike denominators?

4. When you add or subtract fractions with different denominators, how do you find a common denominator?

Think, Pair, Share

Have students jot down their own responses to questions, discuss their responses with a partner (who was not in their station group), and then discuss as a whole class.

Suggested Appropriate Responses

1. fraction circles, strips of paper, paper and pencil, etc.

2. Add the numerators and keep the same denominator.

3. Write each fraction as an equivalent fraction so that the equivalent fractions have common denominators. To find the sum, add the numerators and keep the same denominator.

4. Find the LCM of the denominators. Alternatively, use the product of the denominators as the common denominator.

Possible Misunderstandings/Mistakes

- Adding or subtracting fractions without first finding a common denominator
- Finding an incorrect common denominator
- Incorrectly converting mixed numbers to improper fractions before adding or subtracting

Number and Operations
Set 3: Adding and Subtracting Fractions

Station 1

At this station, you will show the sum of two fractions using fraction circles.

Work together using fraction circles to show the first fraction in each sum. Then add more pieces to the circle to show the second fraction in the sum.

Work with other students to make sure each fraction is shown correctly.

Now find the sum of the two fractions by deciding what part of a full circle is shown. (*Hint:* It is sometimes helpful to place fraction pieces on top of your model to "measure" the amount of a full circle that is shown.)

1. $\dfrac{1}{5} + \dfrac{2}{5}$ = _____

2. $\dfrac{1}{10} + \dfrac{3}{10}$ = _____

3. $\dfrac{1}{6} + \dfrac{1}{3}$ = _____

4. $\dfrac{1}{2} + \dfrac{1}{3}$ = _____

5. $\dfrac{3}{4} + \dfrac{1}{8}$ = _____

6. $\dfrac{3}{8} + \dfrac{5}{8}$ = _____

After showing the two fractions in each sum, how did you decide what part of a complete circle was shown? Describe at least two strategies you used.

_____

_____

_____

_____

_____

Number and Operations
Set 3: Adding and Subtracting Fractions

Station 2

You will use strips of paper to help you subtract fractions.

Fold a strip of paper in half four times, without opening it between folds. Open the strip of paper. You should have 16 equal sections.

To solve the problem $\frac{3}{4} - \frac{3}{8}$, first use a yellow highlighter to color $\frac{3}{4}$ of the strip. Work together to make sure the correct number of sections are colored.

Then use a blue highlighter to color $\frac{3}{8}$ of the strip. (All the sections that you color blue should be within the yellow section.) Work together to make sure the correct number of sections are colored.

The remaining yellow sections show the answer to the subtraction problem. What fraction do the remaining yellow sections show?

_____

Use the above method to solve these problems.

1. $\frac{1}{2} - \frac{1}{16} =$ _____

2. $\frac{3}{4} - \frac{1}{2} =$ _____

3. $\frac{7}{8} - \frac{1}{4} =$ _____

4. $\frac{15}{16} - \frac{3}{4} =$ _____

5. $\frac{5}{8} - \frac{3}{16} =$ _____

Number and Operations
Set 3: Adding and Subtracting Fractions

Station 3
You will be given two sets of cards labeled "A" and "B." Choose one A card and one B card.

Turn the cards over. You will subtract the fraction on card B from the fraction on card A. Write the subtraction problem below.

_____

Work as a group to subtract the fractions. Everyone in the group should agree on your answer. Write the answer below.

_____

Put the cards back. Mix up the cards. Repeat the above process three more times.

   Subtraction problem: _____

   Answer: _____

   Subtraction problem: _____

   Answer: _____

   Subtraction problem: _____

   Answer: _____

Explain the strategies you used to solve the subtraction problems.

_____

_____

_____

_____

_____

Number and Operations
Set 3: Adding and Subtracting Fractions

Station 4

You will need a number cube for this activity. Use the number cube to create fractions.

Roll the number cube 4 times. Write the numbers in the 4 boxes below to create two fractions.

You will add the two fractions. Write the addition problem below.

_____

Work with other students to add the fractions. When everyone agrees on the answer, write it below.

_____

Repeat the process.

Addition problem: _____

Answer: _____

Explain the strategies you used to add the fractions.

_____

_____

_____

_____

Number and Operations

Set 4: Multiplying and Dividing Fractions

Goal: To provide students with opportunities to develop concepts and skills related to multiplication and division

NCTM Standards, Grades 6–8

Number and Operations

Compute fluently and make reasonable estimates: select appropriate methods and tools for computing with fractions and decimals from among mental computation, estimation, calculators or computers, and paper and pencil, depending on the situation, and apply the selected methods.

Student Activities Overview and Answer Key

Station 1

Students model the product of two fractions using 10-by-10 grids. To do so, students model one fraction in the product by shading columns and the other by shading rows. The overlapping region represents the product of the fractions. After working through a step-by-step example of the method, students work together to use the method to find additional products.

Answers: $3/10 \times 1/2 = 3/20$

1. $1/50$; 2. $1/5$; 3. $1/25$; 4. $1/100$

Station 2

Students use rectangular strips of paper to help them divide fractions. They first fold the strips into sixteen equal sections. Given a division problem, students work together to model the first fraction (the dividend) by coloring sections of the strip yellow. Within the yellow section of the strip, students color sections blue to model the divisor. The number of times the blue section would fit inside the yellow section models the answer to the division problem.

Answers: $3/8 \div 1/16 = 6$

1. 4; 2. 12; 3. 4; 4. 2; 5. 2

Station 3

Students are given two sets of cards with fractions written on them. They choose one card from each set and work together to find the quotient of the fractions. At the end of the activity, students discuss the strategies they used to divide the fractions.

Answers: Answers will depend upon the cards that are chosen.

Possible strategies: Divide the fractions by changing the operation to multiplication and changing the divisor to its reciprocal. Then multiply. Simplify the answer.

Station 4

Students are given cards with fractions or mixed numbers written on them. They use the cards and a penny to create various expressions involving multiplication and division of fractions and mixed numbers. Students work together to perform the required operation and simplify the resulting fraction.

Answers: Answers will depend upon the numbers that are chosen.

Materials List/Set Up

Station 1   red and blue pencils

Station 2   yellow and blue highlighters
rectangular strips of paper, about one-inch wide and one-foot long (These may be cut from sheets of paper, or you can use strips of adding-machine tape.)

Station 3   10 index cards, prepared as follows:

- set of 5 cards, labeled "A" on the back with the following fractions written on the front: $3/4$, $4/5$, $5/6$, $15/16$, $5/12$

- set of 5 cards, labeled "B" on the back with the following fractions written on the front: $1/6$, $1/3$, $5/8$, $1/4$, $2/5$

- Place the cards face-down at the station, so that only the backs (A or B) are visible.

Station 4   penny
8 index cards with the following fractions or mixed numbers written on them:
$1/5$, $3/3$, $3/4$, $5/8$, $4/5$, $1\ 1/2$, $1\ 2/3$, $2\ 3/4$

Discussion Guide

To support students in reflecting on the activities and to gather some formative information about student learning, use the following prompts to facilitate a class discussion to "debrief" the station activities.

Prompts/Questions

1. How do you multiply two fractions?

2. How do you multiply two fractions if one of them is a mixed number?

3. How do you divide two fractions?

4. What are some different models, drawings, or tools that you can use to help you multiply and divide fractions?

Think, Pair, Share

Have students jot down their own responses to questions, then discuss with a partner (who was not in their station group), then discuss as a whole class.

Suggested Appropriate Responses

1. Multiply the numerators and multiply the denominators.

2. First change the mixed number to an improper fraction. Then multiply as usual.

3. Change the operation to multiplication and change the divisor to its reciprocal. Then multiply the fractions as usual and simplify the result.

4. 10-by-10 grids, strips of paper, calculators, etc.

Possible Misunderstandings/Mistakes

- Forgetting to take the reciprocal of the divisor when converting a division problem to a multiplication problem

- Not recognizing that whole numbers may be written as fractions in products or quotients by writing them with a denominator of 1

- Incorrectly converting a mixed number to an improper fraction or vice versa

Number and Operations
Set 4: Multiplying and Dividing Fractions

Station 1

Each 10-by-10 grid represents 1. Each column of the grid represents $\frac{1}{10}$. Each row also represents $\frac{1}{10}$.

You can use the grids to multiply fractions. To multiply $\frac{3}{10} \times \frac{1}{2}$, first use a red pencil to shade columns that show $\frac{3}{10}$ of the grid.

Next, use a blue pencil to shade rows that show $\frac{1}{2}$ of the grid.

Work with other students to figure out what fraction of the entire grid is shaded both red and blue. This is the product. Write it below.

_____

Work with other students and use this method to find each product.

1. $\frac{1}{5} \times \frac{1}{10} = $ _____

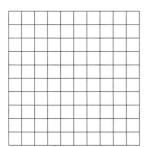

3. $\frac{1}{10} \times \frac{2}{5} = $ _____

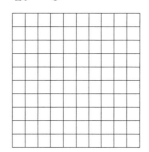

2. $\frac{1}{2} \times \frac{2}{5} = $ _____

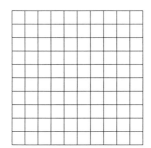

4. $\frac{1}{10} \times \frac{1}{10} = $ _____

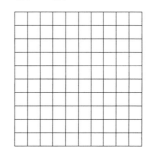

NAME: _____

Number and Operations
Set 4: Multiplying and Dividing Fractions

Station 2

You will use strips of paper to help you divide fractions. Fold a strip of paper in half four times, without opening it between folds. Open the strip of paper. You should have 16 equal sections.

To solve the problem $\dfrac{3}{8} \div \dfrac{1}{16}$, first use a yellow highlighter to color $\dfrac{3}{8}$ of the strip. Work together to make sure the correct number of sections are colored.

Then use a blue highlighter to color $\dfrac{1}{16}$ of the strip. (All the sections that you color blue should be within the yellow section.) Work together to make sure the correct number of sections are colored.

Find the number of times the blue section would fit inside the yellow section. This is the quotient. Write it below.

_____

Use the above method to solve these problems.

1. $\dfrac{1}{2} \div \dfrac{1}{8} =$ _____

2. $\dfrac{3}{4} \div \dfrac{1}{16} =$ _____

3. $1 \div \dfrac{1}{4} =$ _____

4. $\dfrac{3}{4} \div \dfrac{3}{8} =$ _____

5. $\dfrac{5}{8} \div \dfrac{5}{16} =$ _____

Number and Operations
Set 4: Multiplying and Dividing Fractions

Station 3

You will be given two sets of cards labeled "A" and "B." Choose one A card and one B card.

Turn the cards over. Divide the fraction on card A by the fraction on card B. Write the division problem below.

_____

Work as a group to divide the fractions. Everyone in the group should agree on your answer. Write the answer below.

_____

Put the cards back. Mix up the cards. Repeat the above process three more times.

Division problem: _____

Answer: _____

Division problem: _____

Answer: _____

Division problem: _____

Answer: _____

Explain the strategies you used to solve the division problems.

_____

_____

_____

_____

_____

Number and Operations
Set 4: Multiplying and Dividing Fractions

Station 4

You will find a set of cards and a penny at this station. You will use these to create multiplication and division problems involving fractions and mixed numbers.

Choose two cards without looking. Turn the cards over. Write the fractions or mixed numbers in the two boxes below. Then flip the penny. If the penny lands heads up, write "×" on the line between the boxes. If it lands tails up, write "÷" on the line.

Work with other students to find the product or quotient of the fractions or mixed numbers. Simplify the answer if possible. When everyone agrees on the answer, write it below.

_____

Repeat the process four more times.

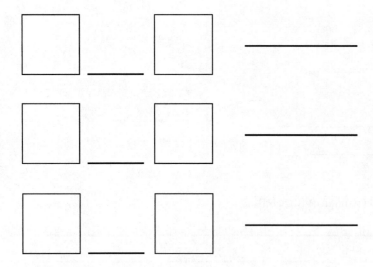

Number and Operations

Set 5: Problem Solving with Fractions, Decimals, and Percents

Goal: To provide opportunities for students to solve problems involving fractions, decimals, and percents

NCTM Standards, Grades 6–8

Number and Operations

Understand numbers, ways of representing numbers, relationships among numbers, and number systems: work flexibly with fractions, decimals, and percents to solve problems.

Student Activities Overview and Answer Key

Station 1

Students work together to solve a problem involving fractions. Students can model the problem using physical objects such as chips or counters. They are encouraged to brainstorm appropriate problem-solving strategies, and to explain their solution process once all students in the group agree upon the solution.

Answers: There were 48 berries in the basket.

Possible strategies: Work backward; model the situation using physical objects.

Possible steps: Work backward to find that there were 8 berries before Ernesto took some, there were 24 berries before Jamal took some, there were 48 berries before Amy took some, and so forth.

Station 2

Students work together to solve a real-world problem involving decimals. After reading the problem, students brainstorm possible problem-solving strategies. Students are encouraged to make sure everyone in the group agrees on the solution. Then students explain their solution process.

Answers: 12 adult tickets and 4 children's tickets

Possible strategies: Guess and check various possibilities; make an organized list of all possible combinations of adult tickets and children's tickets.

Possible steps: Try various combinations of adult and children's tickets that add up to a total of 16 tickets. Find the total cost of the tickets. Adjust the guesses to find a combination for which the total cost is $88.

Station 3

Students work together to solve a problem involving fractions. The problem lends itself especially well to making a table and looking for a pattern. Students can also model the problem using physical objects. Students brainstorm possible strategies, solve the problem, and explain the steps of their solution method.

Answers: At Stage 10, $1/11$ of the tiles are white and $10/11$ of the tiles are black.

Possible strategies: Make a table showing the number of the stage and the corresponding fraction of white tiles and black tiles; look for a pattern.

Possible steps: Look for patterns in the table. Extend the table to Stage 10. Alternatively, notice that the fraction of white tiles at Stage n is $\dfrac{1}{n+1}$, and recognize that the fraction of white tiles plus the fraction of black tiles must equal 1.

Station 4

Students are given problem about distances along a highway involving percents. Students work together to brainstorm strategies they can use to solve the problem. This problem lends itself well to drawing a diagram. After solving the problem, students explain the steps of their solution.

Answers: The distance is 66 miles.

Possible strategies: Draw a diagram.

Possible steps: The distance from Midville to Springfield is 30 miles. The distance from Linwood to Center City is 24 miles. The distance from Springfield to Linwood is $120 - (30 + 24) = 66$ miles.

Materials List/Set Up
Station 1	about 50 small objects such as chips, counters, beans, or pennies
Station 2	none
Station 3	small black and white tiles, or small square pieces of paper in two colors
Station 4	none

Discussion Guide

To support students in reflecting on the activities, and to gather formative information about student learning, use the following prompts to facilitate a class discussion to "debrief" the station activities.

Prompts/Questions

1. What are some different problem-solving strategies you can use to help you solve real-world problems?

2. How do you know when you should use multiplication to solve a problem?

3. How do you know when you should use division to solve a problem?

4. How can you check your answer to a real-world problem?

Think, Pair, Share

Have students jot down their own responses to questions, discuss their responses with a partner (who was not in their station group), and then discuss as a whole class.

Suggested Appropriate Responses

1. Make a table, guess and check, look for a pattern, work backward, draw a diagram, use physical objects to model the problem, and so forth.

2. You use multiplication when equal groups of objects are being put together.

3. You use division when a set of objects is being separated into equal groups.

4. Reread the problem using the answer in place of the unknown quantity or quantities. Check to see if the numbers work out correctly throughout the problem.

Possible Misunderstandings/Mistakes

- Incorrectly multiplying or dividing fractions
- Incorrectly identifying, describing, or extending a number pattern in a list or table
- Drawing an inaccurate diagram to help solve a problem

Number and Operations
Set 5: Problem Solving with Fractions, Decimals, and Percents

Station 1

You will find some chips or counters at this station. Use them to help you solve this problem.

Three friends shared a basket of berries. Amy took $\frac{1}{2}$ of the berries. Then Jamal took $\frac{1}{3}$ of the remaining berries. Then Ernesto took $\frac{1}{4}$ of the remaining berries. There were 2 berries left in the basket. How many berries were in the basket at the beginning?

Discuss the problem with other students. Brainstorm strategies you might use to solve the problem. Write the strategies below.

_____
_____
_____
_____

Solve the problem. When everyone agrees on the answer, write it below.

_____

Explain the steps you used to solve the problem.

_____
_____
_____
_____
_____
_____

Number and Operations
Set 5: Problem Solving with Fractions, Decimals, and Percents

Station 2

At this station you will work with other students to solve this real-world problem.

Alicia sells tickets to a school play. The tickets cost $6.50 for adults and $2.50 for children. She sells a total of 16 tickets and collects a total of $88. How many of each type of ticket does Alicia sell?

Discuss the problem with other students. Brainstorm strategies you might use to solve the problem. Write the strategies below.

_____

_____

_____

_____

Solve the problem. When everyone agrees on the answer, write it below.

_____

Explain the steps you used to solve the problem.

_____

_____

_____

_____

_____

_____

Number and Operations
Set 5: Problem Solving with Fractions, Decimals, and Percents

Station 3

You will find a set of tiles at this station. Use them to help you solve this problem.

Mei is making a pattern of black and while tiles for the floor of a bathroom. The figures show the pattern at several stages. At Stage 10, what fraction of the tiles are white? What fraction of the tiles are black?

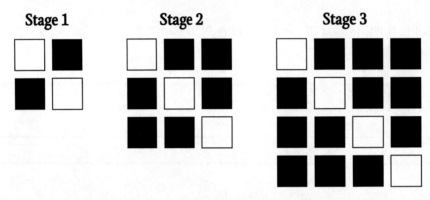

Stage 1   Stage 2   Stage 3

Discuss the problem with other students. Brainstorm strategies you might use to solve the problem. Write the strategies below.

_____

_____

_____

_____

Solve the problem. When everyone agrees on the answer, write it below.

_____

Explain the steps you used to solve the problem.

_____

_____

_____

_____

_____

_____

Number and Operations
Set 5: Problem Solving with Fractions, Decimals, and Percents

Station 4

At this station you will work with other students to solve this real-world problem.

The distance along a straight highway from Midville to Center City is 120 miles. The distance along the highway from Midville to Springfield is 25 percent of the distance from Midville to Center City. The distance along the highway from Linwood to Center City is 20 percent of the distance from Midville to Center City. What is the distance from Springfield to Linwood?

Discuss the problem with other students. Brainstorm strategies you might use to solve the problem. Write the strategies below.

_____

_____

_____

_____

Solve the problem. When everyone agrees on the answer, write it below.

_____

Explain the steps you used to solve the problem.

_____

_____

_____

_____

_____

_____

Geometry and Measurement

Set 1: Appropriate Units of Measurement

Goal: To provide opportunities for students to develop concepts and skills related to using appropriate units for measurement

NCTM Standards, Grades 6–8

Geometry and Measurement

Understand measurable attributes of objects and the units, systems, and processes of measurement: understand, select, and use units of appropriate size and type to measure angles, perimeter, area, surface area, and volume.

Apply appropriate techniques, tools, and formulas to determine measurements: select and apply techniques and tools to accurately find length, area, volume, and angle measures to appropriate levels of precision.

Student Activities Overview and Answer Key

Station 1

At this station, students will measure a variety of rectangular prisms and find the volume of these objects. Students will use appropriate measuring techniques and units.

Answers: Answers will vary; inches; square inches; cubic inches; it depends on how many times you multiply inches

Station 2

Students will measure the length and width of the classroom to determine its area. They will explain their use of appropriate units.

Answers: Answers will vary; answers will vary; feet/yards because inches are too small; answers will vary; use the yard stick, choose units that were large, etc.

Station 3

Students find the approximate volume of their hand. They do this by finding the volume of six parts of the hand and adding that together. They then comment on their findings.

Answers: Answers will vary; cubic inches/centimeters, etc.; fingers/palms are not exactly rectangular prisms; answers may vary

Station 4

Students choose five objects in the classroom. They find the perimeter of these objects then reflect on the units that they chose to measure the objects in.

Answers: Answers will vary.

Materials List/Set Up

Station 1	variety of items that are rectangular prisms; rulers
Station 2	yard stick; ruler
Station 3	rulers; calculators—enough for all group members
Station 4	yard stick; ruler

Discussion Guide

To support students in reflecting on the activities, and to gather formative information about student learning, use the following prompts to facilitate a class discussion to "debrief" the station activities.

Prompts/Questions

1. What is the difference between the units used in perimeter, area, and volume?

2. If you are measuring the side of an object using inches, and find that the length of the object is between 13 inches and 13 $\frac{1}{16}$, but closer to 13 $\frac{1}{16}$ inch, what do you use as the measure?

3. If you have two objects and one has a greater perimeter, does it have to have greater areas as well?

4. How is finding perimeter different from finding area?

Think, Pair, Share

Have students jot down their own responses to questions, discuss their responses with a partner (who was not in their station group), and then discuss as a whole class.

Suggested Appropriate Responses

1. perimeter = units, area = square units, volume = cubic units

2. The answer is 13 $\frac{1}{16}$ inches because we round to the nearest $\frac{1}{16}$ if that is the most precise our ruler can get.

3. no (e.g., 1×14 and 5×5)

4. We add all the lengths of the sides for perimeter but multiply when finding area.

Possible Misunderstandings/Mistakes

- Getting confused between $\frac{1}{4}$, $\frac{1}{8}$, and $\frac{1}{16}$ of an inch
- Difficulty measuring the hand
- Trouble deciding when it is better to use inches, feet, or yards

Geometry and Measurement
Set 1: Appropriate Units of Measurement

Station 1

At this station, you will find a variety of objects and rulers. You will be finding the volume of these objects.

Each student should choose a different object. You will measure the length, width, and height of your object. Record the group data in the table below.

Group Member	Object Chosen	Length (in.)	Width (in.)	Height (in.)	Total Volume

Which object has the greatest volume? _____

What units did you write the volume in? _____

How did you know what units to use? _____

_____

_____

Geometry and Measurement
Set 1: Appropriate Units of Measurement

Station 2

Your goal at this station is to find the area of the room. You have a ruler and a yard stick to help you with this.

First measure the length and the width of the room and write them below.

Length (include units): _____

Width (include units): _____

Why did you choose the units you did? _____

_____

_____

What is the area of the room? (include units) _____

What was your strategy for measuring the room and deciding which units to use?

_____

_____

_____

_____

Geometry and Measurement
Set 1: Appropriate Units of Measurement

Station 3

At this station, you will be finding the approximate volume of your hand. You will be using rulers and a calculator.

For this activity, think of your hand as six different rectangular prisms—the five fingers and your palm. Measure the length, width, and height of each of the six parts of the hand. It may be helpful to work with a group member for the measuring section of this activity.

Part of Hand	Length	Width	Height	Volume
Pinky finger				
Ring finger				
Middle finger				
Index finger				
Thumb				
Palm				

What is the approximate volume of your hand? _____

What units did you use? Why? _____

_____

_____

Why might your results not be exact? _____

_____

_____

Which finger has the most volume? Does this surprise you? _____

_____

_____

Geometry and Measurement
Set 1: Appropriate Units of Measurement

Station 4

At this station, you will find the perimeter of five different objects in the classroom. You will have a ruler and a yard stick to assist you.

First, choose five different objects to find the perimeter of (e.g., desk, room, etc).

Object	Length	Width	Perimeter

What object had the largest perimeter? _____

What units did you use? Were they always the same? _____

_____

_____

How did you decide what units to use? _____

_____

_____

Geometry and Measurement

Set 2: Volume and Surface Area

Goal: To provide opportunities for students to develop concepts and skills related to determining the volume and surface area of solid figures

NCTM Standards Grades 6–8

Geometry and Measurement

Apply appropriate techniques, tools, and formulas to determine measurements: develop strategies to determine the surface area and volume of selected prisms, pyramids, and cylinders

Student Activities Overview and Answer Key

Station 1

Students will trace all six sides of a rectangular prism. They will then add up the areas to determine the surface area. They will reflect on what they did and see if they need to find the area of all the sides.

Answers: Answers will vary. Yes, opposite sides had the same area; we could find the area of the three unique sides, then multiply that number by 2.

Station 2

Students use 1 cubic inch blocks and a rectangular container to determine the formula for finding the volume of a rectangular prism. They do this by filling the container with the cubes, and comparing that to multiplying the length × the width × the height.

Answers: 2 in; 4 in; 8 in; 64; 64 cubic inches; 64 cubic inches; Vol = *lwh*

Station 3

Students discover the formula for the volume of a pyramid. They find the volume of a cube, then pour water from the pyramid until the cube is full. They make observations and determine the formula.

Answers: Answers will vary; answers will vary; answers will vary; answers will vary; 3; divide the volume of the cube by 3; $V = (1/3)lwh$

Station 4

Students explore how to find the surface area of a cylinder. They do this by tracing one onto grid paper and finding the area of the individual pieces. They then consider how these pieces are related and finally develop a formula.

Answers: A rectangle; the circumference; answers will vary; the two circles plus the rectangle = $2\pi r^2 + 2\pi r h$

Materials List/Set Up

Station 1 a rectangular prism

Station 2 a container that is 2in × 4in × 8in; at least 64 blocks that are all 1 cubic inch; a ruler

Station 3 pyramid and cube (same dimensions as pyramid) such that water can be poured in and out of them; water; ruler

Station 4 paper; scissors; a cylinder

Discussion Guide

To support students in reflecting on the activities, and to gather formative information about student learning, use the following prompts to facilitate a class discussion to "debrief" the station activities.

Prompts/Questions

1. What is a strategy for finding surface area if we do not know the formula?

2. What is an example of when you would want to know surface area in the real-world?

3. When it is important to know the volume of a rectangular prism in the real-world?

4. What is a strategy for finding volume if we do not know the formula?

Think, Pair, Share

Have students jot down their own responses to questions, then discuss with a partner (who was not in their station group), and then discuss as a whole class.

Suggested Appropriate Responses

1. Trace the figure onto grid paper and use that to help.

2. Many possibilities—when making a product, use the least amount of material for the most amount of volume

3. Many possibilities—when going on vacation: a suitcase

4. Compare the volume of the figure to the volume of something we know

Possible Misunderstandings/Mistakes

- Confusion regarding a cylinder being made up of two circles and a rectangle
- Trouble with the concept of π
- Trouble transferring water from the pyramid to the cube—spilling, which skews the results, etc.

Geometry and Measurement
Set 2: Volume and Surface Area

Station 1

At this station, you will be finding the surface area of a rectangular prism.

Place a rectangular prism on the grid below. Trace each side.

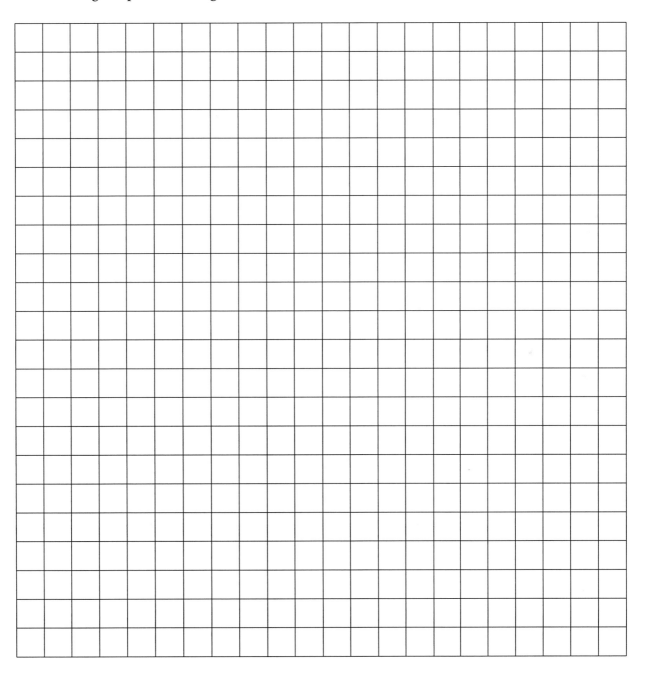

Geometry and Measurement
Set 2: Volume and Surface Area

Station 1, *continued*

If you sum the areas of all six sides, what total surface area do you get? _____

Were any two sides the same? Explain. _____

_____

_____

Is there an easier way to find the surface area than measuring all the sides? Explain.

_____

_____

_____

Geometry and Measurement
Set 2: Volume and Surface Area

Station 2

At this station you will use inch square cubes and a container to help you develop a formula for determining volume of a rectangular prism.

Measure the length, width, and height of the container.

        Length = _____

        Width = _____

        Height = _____

Now place as many cubes inside the container as possible.

How many cubes did you fit inside the container? _____

If one cube is 1 cubic inch, what is the volume of the container? _____

What is the length? the width? the height?

        Length = _____

        Width = _____

        Height = _____

What do you notice about this number? _____

_____

_____

How can you find the volume of a rectangular container? _____

_____

_____

Geometry and Measurement
Set 2: Volume and Surface Area

Station 3

At this station, you will see a pyramid container and a cube container. You will be using these to help determine a formula for finding the volume of a pyramid.

First find the volume of the cube. _____

Find the dimensions of the pyramid.

Length = _____

Width = _____

Height = _____

Now fill the pyramid with water. Pour it into the cube. Repeat until full.

How many pyramids full of water fit into the cube? _____

If you know the volume of the cube, how can you find the volume of the pyramid?

_____

_____

_____

What is the formula for finding the volume of a pyramid? _____

Geometry and Measurement
Set 2: Volume and Surface Area

Station 4

At this station, you will be finding the surface area of a cylinder.

Place a cylinder on the grid below. Trace the top and bottom.

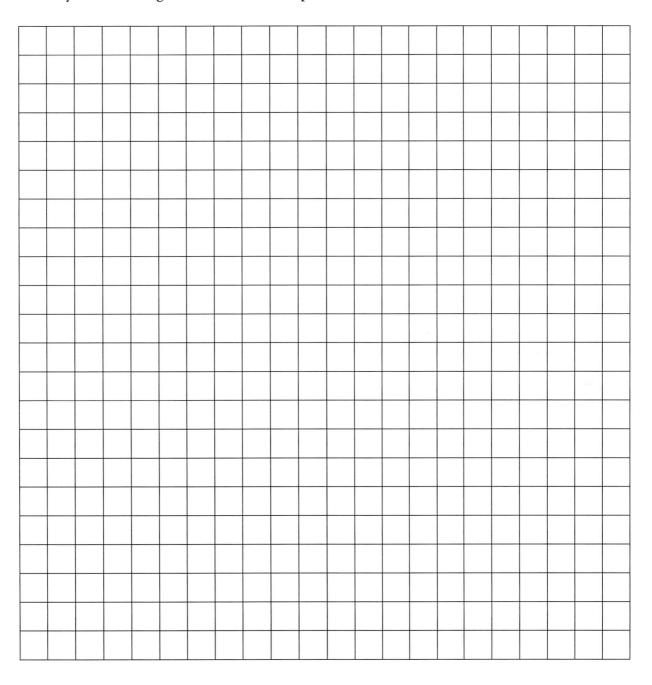

continued

Geometry and Measurement
Set 2: Volume and Surface Area

Station 4, *continued*

Now take a sheet of paper and cut it so it is the same height as the cylinder. Wrap it around the cylinder, marking off exactly where it begins to overlap. Cut the paper so it fits exactly around the side of the cylinder in both height and width. Trace this onto the grid as well.

What shape did you just trace onto the paper? _____

The width of this rectangle is the _____ of the end circles.

Find the area of the three shapes on your grid. What is the total sum? _____

Can you come up with a formula for finding the surface area of a cylinder? Explain.

_____

_____

_____

Geometry and Measurement

Set 3: Ratio, Proportion, and Scale

Goal: To provide opportunities for students to develop concepts and skills related to demonstrating the relationship between similar plane figures using ratio, proportion, and scale factor

NCTM Standards, Grades 6–8

Geometry and Measurement

Apply appropriate techniques, tools, and formulas to determine measurements: solve problems involving scale factors, using ratio and proportion.

Student Activities Overview and Answer Key

Station 1

Students construct their own triangles with given angle measurements. They then compare their triangles and explore what similarity means with respect to triangles.

Answers: The ratios are the same; yes, because the ratios are the same and the angles are the same

Station 2

Students will use a drawing of a room and the scale to determine the dimensions of an actual room. They then explain their strategy for successfully completing the activity.

Answers: $1\ ^9/_{16}$ in $\times$ $2\ ^1/_8$ in; 15.625 ft $\times$ 21.25 ft; many possible answers—set up a proportion, etc.

Station 3

Students apply scale factor and draw a larger picture using this concept. They then reflect on their strategies for successfully completing the task.

Answers: Answers will vary—setting up proportions, etc.

Station 4

Students have two quadrilaterals that they work with to determine scale factor. They measure each corresponding side and find the ratio. They then reflect on the task.

Answers: $5/4$; that is the ratio you need to multiply the length of the sides of the smaller quadrilateral by in order to get the larger quadrilateral

Materials List/Set Up

Station 1   pencils, protractors, rulers, and calculators for all group members

Station 2   rulers and calculators for all group members

Station 3   protractors, rulers, and calculators for all group members

Station 4   protractors, rulers, and calculators for all group members

Discussion Guide

To support students in reflecting on the activities, and to gather formative information about student learning, use the following prompts to facilitate a class discussion to "debrief" the station activities.

Prompts/Questions

1. If you had a figure and wanted to construct a similar figure, what information would be important to know?

2. When is scale factor important in real life?

3. How do proportions help you when working with scale factor?

4. When do we use similar figures in real life?

Think, Pair, Share

Have students jot down their own responses to questions, discuss their responses with a partner (who was not in their station group), and then discuss as a whole class.

Suggested Appropriate Responses

1. the angles, proportion of the sides

2. Many possibilities—in blueprints

3. We can set up a proportion to find the new dimensions if we know the scale factor, or we can set up a proportion to find the scale factor if we know our new dimensions.

4. Many possibilities—different size plates—dinner plate vs. bread plate, etc.

Possible Misunderstandings/Mistakes

- Not accurately measuring angles, which will change the ratios
- Not accurately measuring with a ruler, which will change the ratios
- Getting caught up in what the state of Georgia looks like when the concept of scale factor is what is important

Geometry and Measurement
Set 3: Ratio, Proportion, and Scale

Station 1

At this station you will find rulers, pencils, calculators, and protractors. You will be using these materials to draw a triangle and make observations.

Each group member should draw a triangle in the space below. The angles of the triangle should be 90°, 30°, and 60°. The 90° angle is angle *A;* the 30° angle is angle *B;* the 60° angle is angle *C.*

Each person should measure the sides of his/her triangle. Compile your data in the table below.

Group Member	Length *AB* (cm)	Length *BC* (cm)	Length *AC* (cm)	*AB/BC*	*BC/AC*	*AB/AC*

What do you notice about the ratio of the sides? _____

_____

_____

Are the triangles similar? How do you know? _____

_____

_____

Geometry and Measurement
Set 3: Ratio, Proportion, and Scale

Station 2

At this station, you will take on the role of an architect. You will find rulers and calculators to help you with this role.

Below is a diagram of a room in a house. You need to figure out the dimensions of the actual room in the house. For every inch in the diagram, the real room is 10 feet.

Hint: You should use your ruler to find the dimensions of the diagram.

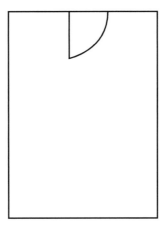

What are the dimensions? _____

What are the dimensions of the actual room? _____

What was your strategy for figuring out the actual dimensions? _____

_____

_____

_____

Geometry and Measurement
Set 3: Ratio, Proportion, and Scale

Station 3

At this station, you will find enough rulers and calculators for all group members.

Your job is to redraw a picture of the continental United States at a scale factor 3 times larger than what it is now.

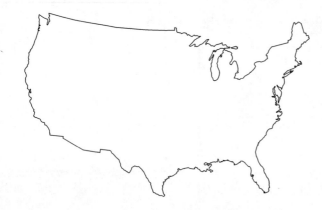

What was your strategy for successfully completing this task? _____

_____

_____

_____

Geometry and Measurement

Station 4

At this station, you will find rulers and calculators to help you find the scale factor between two figures.

Look at the two figures below. They are similar. Your job is to determine the scale factor used to go from the first figure to the second figure.

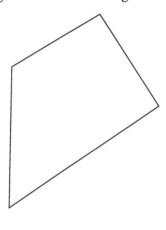 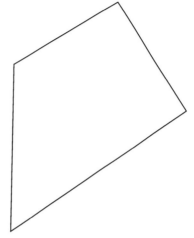

First, measure the corresponding sides.

Side	Quadrilateral 1	Quadrilateral 2
1		
2		
3		
4		

Use this information to determine the scale factor.

What is the scale factor? _____

How do you know? _____

_____

_____

Geometry and Measurement

Set 4: Visualizing Solid Figures

Goal: To provide opportunities for students to develop concepts and skills related to visualizing solid figures

NCTM Standards, Grades 6–8

Geometry and Measurement

Use visualization, spatial reasoning, and geometric modeling to solve problems: draw geometric objects with specified properties, such as side lengths or angle measures; use two-dimensional representations of three-dimensional objects to visualize and solve problems such as those involving surface area and volume.

Student Activities Overview and Answer Key

Station 1

Students will be drawing figures made from connecting cubes from different perspectives. They will then comment on their strategies for doing this.

Answers: Answers will vary; no—some are hidden by other blocks in some perspectives

Station 2

Students look at different possible nets for a cube. They decide whether or not they are, in fact, nets and then draw a net of their own.

Answers: yes; no; yes; yes; no; answers will vary

Station 3

Students will use connecting cubes to construct a figure based on a drawing. They then create and draw their own figure. Finally, students reflect on their strategies for completing the task.

Answers: 15; yes—they are hidden behind other cubes; answers will vary

Station 4

Students draw nets for two different cylinders. They then compare and contrast these nets, and determine which cylinder has a greater surface area.

Answers: The rectangle is the same for both; the circles on the ends are different for both; the cylinder with the end circle whose circumference is 11

Materials List/Set Up

Station 1	3 different figures made up of 10 or fewer connecting cubes
Station 2	none
Station 3	25 connecting cubes
Station 4	tape; two 8 $\frac{1}{2}$" × 11" sheets of paper

Discussion Guide

To support students in reflecting on the activities and to gather some formative information about student learning, use the following prompts to facilitate a class discussion to "debrief" the station activities.

Prompts/Questions

1. When in real life do we have 2-D drawings of 3-D objects?

2. How many nets can you think of for a cube?

3. When would having a drawing on isometric dot paper be more useful than having a net for an object?

4. How can nets help us find surface area?

Think, Pair, Share

Have students jot down their own responses to questions, discuss their responses with a partner (who was not in their station group), and then discuss as a whole class.

Suggested Appropriate Responses

1. Many examples—topological maps, architectural elevations, etc.

2. 11

3. when you want to see depth

4. If we find the area of the net, we know the surface area of the object.

Possible Misunderstandings/Mistakes

- Not being able to visualize the nets—thinking they do or do not work when the reverse is true
- Not correctly drawing the figure they build
- Not being able to build the figure from the isometric dot paper—being confused by the perspective
- Confusing the front, side, and top drawings

Geometry and Measurement
Set 4: Visualizing Solid Figures

Station 1

At this station, you will find connecting cubes that have been made into three different figures. You will use these figures to draw from different perspectives.

Split your group up into three subgroups and give each group one figure.

Draw the front perspective, the top perspective, and the right side perspective of the figures on the grid below. Be sure to label which perspective you are drawing.

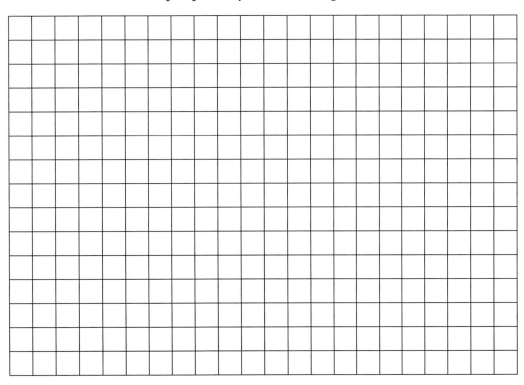

What were your strategies for this activity? _____

_____

_____

Can you see all the blocks from all perspectives? Explain. _____

_____

_____

Geometry and Measurement
Set 4: Visualizing Solid Figures

Station 2

At this station you will be working with nets.

Below are possible nets for cubes.

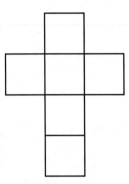

Could you fold this along the lines to make a cube? _____

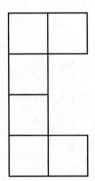

Could you fold this along the lines to make a cube? _____

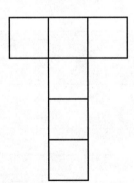

Could you fold this along the lines to make a cube? _____

continued

Geometry and Measurement
Set 4: Visualizing Solid Figures

Station 2, *continued*

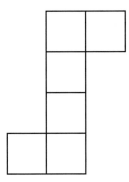

Could you fold this along the lines to make a cube? _____

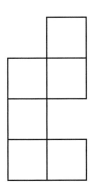

Could you fold this along the lines to make a cube? _____

Draw a net of your own that can be folded along the lines to make a cube.

Geometry and Measurement
Set 4: Visualizing Solid Figures

Station 3

At this station, you will find 25 connecting cubes. You will use these to build a figure.

Below is a figure drawn on isometric dot paper. Make this figure using connecting cubes.

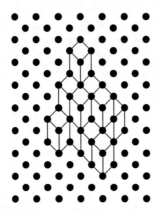

How many cubes did you use? _____

Are there any cubes that you cannot see in the drawing? Explain. _____

_____

_____

Now make a figure of your own out of the connecting cubes. Draw your figure.

What were your strategies for completing this task? _____

_____

_____

_____

Geometry and Measurement
Set 4: Visualizing Solid Figures

Station 4

At this station, you will find two sheets of paper and tape. You will use these to draw nets for two cylinders.

Tape one sheet of paper into a cylinder so that 11 inches is the circumference. Tape the other sheet of paper into a cylinder so that 11 inches is the height.

Draw the net for each cylinder on the grid below. You may draw the net to be $\frac{1}{4}$ of the actual size. Be sure to label which sheets of paper belong to which cylinder.

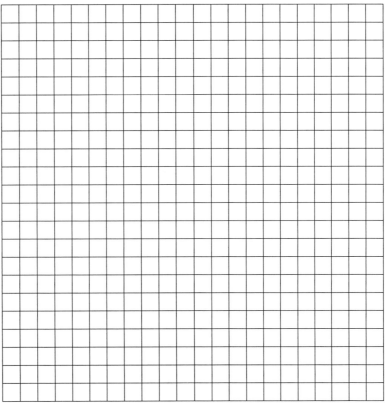

What similarities do you see between the two nets?_____

_____

_____

What differences do you see between the two nets? _____

_____

_____

Which cylinder has a greater surface area? _____

Geometry and Measurement

Set 5: Problem Solving with Volume, Surface Area, and Scale

Goal: To provide opportunities for students to develop concepts and skills related to problem solving involving volume, surface area, and scaling

NCTM Standards, Grades 6–8

Geometry and Measurement

Apply appropriate techniques, tools, and formulas to determine measurements: develop strategies to determine the surface area and volume of selected prisms, pyramids, and cylinders; solve problems involving scale factors, using ratio and proportion.

Student Activities Overview and Answer Key

Station 1

Students construct a cylinder out of a sheet of printer paper. They explore whether the way they choose to roll the paper affects the volume. They experiment and use calculations to answer this question.

Answers: Answers will vary; the cylinder with the 11-inch circumference; 257.1 cubic inches; 198.7 cubic inches; the cylinder with the 11-inch circumference

Station 2

At this station, students use the formula for the volume of a cone to determine how many ice cream cones could fit into a cooler. They explain how they do this, and are instructed to be careful with their units.

Answers: πr^2; 12.57 cubic inches; 12 cubic feet; 137 cones; you need to change feet to inches or inches to feet so you are working in the same unit, then you can divide the volume of a cone into the volume of the cooler

Station 3

Students use their knowledge of surface area to determine the surface area of a bird house. They need to be able to find the surface area of a rectangular prism and of a cylinder. Finally, they explain how they arrived at their solution.

Answers: 128 square inches; 6.28 square inches—the area of the 2 circles; 6.28 square inches—the distance around the inside of the circle; 128 square inches; answers will vary

Station 4

Students use a photo and a picture frame to explore scale factor. They need to determine the scale factor necessary to enlarge the photo so that it fits in the frame.

Answers: Finding the dimensions to the photo and frame; answers will vary; answers will vary; answers will vary; find the ratio between the photo and the frame; ask them to enlarge the photo by the scale factor

Materials List/Set Up

Station 1	2 sheets of printer paper, tape, and mini marshmallows
Station 2	a calculator
Station 3	calculators for all group members
Station 4	calculators and rulers for all group members a small picture and a larger picture frame (they must be similar)

Discussion Guide

To support students in reflecting on the activities, and to gather formative information about student learning, use the following prompts to facilitate a class discussion to "debrief" the station activities.

Prompts/Questions

1. Why does the way you roll a rectangle affect the volume of the cylinder you create?

2. What is a real-life example of when knowing the volume of an object is important?

3. What is a real-life example of when knowing the surface area of an object is important?

4. What is a real-life example of when scale factor is important?

Think, Pair, Share

Have students jot down their own responses to questions, discuss their responses with a partner (who was not in their station group), and then discuss as a whole class.

Suggested Appropriate Responses

1. One way, the length of the rectangle is the height of the cylinder, and the other way, it is the circumference of the cylinder.

2. Many examples—a can of soup, storage container, etc.

3. Many examples—when painting, wrapping, covering, etc.

4. Many examples—blueprints, maps, plans, etc.

Possible Misunderstandings/Mistakes

- Not being able to find radius when given the circumference
- Having trouble figuring out the surface area of the hole in the bird house
- Measuring the outside of the picture frame instead of the area that the picture will go in

Geometry and Measurement
Set 5: Problem Solving with Volume, Surface Area, and Scale

Station 1

At this station, you will find two sheets of paper, tape, and mini marshmallows. Each sheet of paper is 8 $\frac{1}{2}$" × 11". You will be using these to investigate volume.

Your task is to answer this question: Which way should you roll the paper into a cylinder to get the greatest volume? with 11 inches as the circumference or the height? Or, does it matter?

Tape one sheet of paper into a cylinder so that 11 inches is the circumference. Tape one sheet of paper into a cylinder so that 11 inches is the height.

Which do you think holds more? _____

Fill both cylinders with mini marshmallows.

Which cylinder holds more? _____

Now calculate the volume of the two cylinders. Write your solution below.

Vol. 11" circumference: _____

Vol. 11" height: _____

Which cylinder has a greater volume? _____

Geometry and Measurement
Set 5: Problem Solving with Volume, Surface Area, and Scale

Station 2

At this station, you will pretend you are an ice cream vender. You have a cooler that is 3 feet × 2 feet × 2 feet. You want to fit in as many ice cream cones as possible. Each ice cream cone is pre-wrapped and has ice cream in it, so you cannot fit them inside one another.

The way to find the volume of a cone is the same as finding the volume of a pyramid—$\left(\frac{1}{3}\right)bh$. In the case of the cone, the base is not found by multiplying the length times the width.

How do we find the area of the base of a cone? _____

_____

_____

Each ice cream cone is 4 inches tall, and the radius is 1 inch.

What is the volume of one ice cream cone? _____

What is the volume of the cooler? _____

How many ice cream cones can you fit in the cooler? (Be careful of your units!)

_____

Explain how you arrived at your solution. _____

_____

_____

_____

Geometry and Measurement
Set 5: Problem Solving with Volume, Surface Area, and Scale

Station 3

Imagine you wanted to build and paint a bird house. You have already completed most of the bird house, but you still have the most challenging piece to worry about: the front piece.

The front of the bird house looks like the drawing below. The radius of the circle is 1 inch. The block of wood is 1 inch by 4 inches by 12 inches.

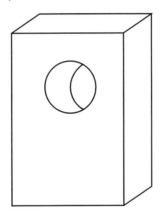

What is the surface area of the block of wood if it did not have the circle in it?

_____

What surface area are you losing because of the missing circle? Explain. _____

_____

_____

What surface area are you gaining because the circle is missing? Explain. _____

_____

_____

What is the overall surface area with the circle cut out? _____

Explain your strategy for figuring this out. _____

_____

_____

_____

Geometry and Measurement
Set 5: Problem Solving with Volume, Surface Area, and Scale

Station 4

At this station, you will find enough calculators and rulers for all group members. You will also find a small picture and a larger picture frame.

You are asked to enlarge the small picture so it will fit into the picture frame.

What is the first step to solving this problem? _____

_____

_____

What are the dimensions of the photo?

      Length = _____

      Width = _____

      Height = _____

What are the dimensions of the frame?

      Length = _____

      Width = _____

      Height = _____

What is the scale factor? _____

How did you arrive at that answer? _____

_____

_____

Now that you have this information, what would you ask for if you went into a photo shop?

_____

_____

_____

Algebra

Set 1: Patterns and Relationships

Goal: To provide opportunities for students to develop concepts and skills related to patterns and relationships

NCTM Standards, Grades 6–8

Algebra

Understand patterns, relations, and functions: represent, analyze, and generalize a variety of patterns with tables, graphs, words, and, when possible, symbolic rules.

Student Activities Overview and Answer Key

Station 1

Students build, extend, and describe a geometric pattern made of toothpicks. They are asked to work together to predict the number of toothpicks that would be needed to build the tenth stage of the pattern, and they are asked to describe a general rule for the pattern.

Answers: 1. 9, 11; 2. 21; 3. Possible answer: Double the number of the stage and add 1. 4. Possible rule: Double the number of the stage and add 1.

Station 2

Students work together to analyze a pattern based on the cost of a gym membership. They work together to extend a table of values and predict the cost of joining the gym for various numbers of months. The emphasis is on recognizing and describing the underlying numerical pattern.

Answers: 1. $92, $104; 2. Possible explanation: Extend the table. For each additional month, the cost increases by $12; 3. 10 months. 4. Possible explanation: Extend the table until the cost is $140 and read the corresponding number of months. 5. Possible description: For each additional month, the cost increases by $12. (Although students are not required to use variables, they may notice that the cost for n months is $20 + 12n$.)

Station 3

Students work together to analyze and extend a pattern that is provided in the form of a table. The table shows the number of baseball cards in a collection from one week to the next. Students are asked to predict the number of cards in future weeks and to describe the pattern.

Answers: 1. 87; 2. Possible explanation: Extend the table to the ninth week by adding 8 cards per week. 3. 14 weeks; 4. Extend the table until the number of cards is greater than 120 and count the number of weeks. 5. Possible description: For each additional week, the number of cards increases by 8. (Although students are not required to use variables, they may notice that the number of cards on week n is $15 + 8n$.)

Station 4

Students build, extend, and describe a pattern based on tables and chairs. To do so, they may use physical objects to model the pattern. Students work together to determine how many chairs are needed for a given number of tables, and then generalize their work by describing a rule for the pattern.

Answers: 1. 10 chairs; 12 chairs; 2. 26 chairs; 3. Possible explanation: Double the number of tables and add 2. 4. Possible rule: Multiply the number of tables by 2 and then add 2.

Materials List/Set Up

Station 1	box of toothpicks
Station 2	none
Station 3	none
Station 4	small square tiles and small round tiles

Discussion Guide

To support students in reflecting on the activities, and to gather formative information about student learning, use the following prompts to facilitate a class discussion to "debrief" the station activities.

Prompts/Questions

1. What are some different tools, objects, or drawings that you can use to help you analyze a pattern?

2. What are some strategies you can use to help extend a number pattern?

3. How can you use a calculator to check your work when you extend a number pattern?

4. Can all the patterns in these activities be extended indefinitely? Why or why not?

Think, Pair, Share

Have students jot down their own responses to questions, discuss their responses with a partner (who was not in their station group), and then discuss as a whole class.

Suggested Appropriate Responses

1. Hands-on manipulatives, drawings of patterns, tables, etc.

2. Look for a pattern in the sequence of numbers. For example, a fixed number may be added each time you go from one number to the next.

3. Use the calculator to check that the repeated addition was performed correctly.

4. Yes. Each pattern is based on adding a fixed number at each stage of the pattern. That process can be continued forever.

Possible Misunderstandings/Mistakes

*   Incorrectly determining the value by which a numerical pattern increases or decreases from one term to the next

*   Assuming that all patterns may be described multiplicatively (i.e., in the form $y = kx$)

*   Incorrectly reproducing a given pattern with manipulatives or with a drawing

Algebra
Set 1: Patterns and Relationships

Station 1

You will find some toothpicks at this station. Use them to help you with this activity.

Here are several stages of a pattern made from toothpicks.

Stage 1   **Stage 2**   **Stage 3**

Work with other students to build the pattern from toothpicks.

Then work together to answer the following questions. When everyone agrees on an answer, write it in the space provided.

1. How many toothpicks do you need to make Stage 4 and Stage 5? _____

2. Predict the number of toothpicks you would need to make Stage 10. _____

3. Explain how you made this prediction. _____

    _____

    _____

    _____

4. Describe a rule for the pattern. That is, if you are given the stage of the pattern, describe a rule for finding the number of toothpicks needed.

    _____

    _____

    _____

Algebra
Set 1: Patterns and Relationships

Station 2

At this station, you will explore a pattern based on prices.

A gym posts this table at the front desk. It shows the cost of joining the gym for different numbers of months.

Number of Months	1	2	3	4	5
Cost of Membership	$32	$44	$56	$68	$80

Work together to answer the following questions. When everyone agrees on an answer, write it in the space provided.

1. Predict the cost of joining the gym for 6 months and for 7 months. _____

2. Explain how you made these predictions. _____

_____

_____

_____

3. Janelle has $140 to spend on a membership to this gym. For how many months can she join?

_____

4. Explain how you found the answer to Question 3. _____

_____

_____

_____

5. Describe the pattern in the table. _____

_____

_____

_____

Algebra
Set 1: Patterns and Relationships

Station 3

At this station, you will explore a pattern based on a table of data.

Diego collects baseball cards. Each week, he buys some new cards. He starts with 23 cards in his collection the first week. The table below shows the number of cards in Diego's collection each week.

Week	1	2	3	4	5
Number of Cards	23	31	39	47	55

Work together to answer the following questions. When everyone agrees on an answer, write it in the space provided.

1. Predict the number of cards in Diego's collection on the 9th week. _____

2. Explain how you made this prediction. _____

   _____

   _____

   _____

3. How many weeks will it take Diego to have more than 120 cards in his collection?

   _____

4. Explain how you found the answer to Question 3. _____

   _____

   _____

   _____

5. Describe the pattern in the table. _____

   _____

   _____

   _____

NAME: _____

Algebra
Set 1: Patterns and Relationships

Station 4

At this station, you will find some tiles that you may use to help you analyze a pattern.

A restaurant has square tables that can be pushed together in a row to seat large groups. The figure below shows the number of chairs (circles) that are needed for various numbers of tables (squares).

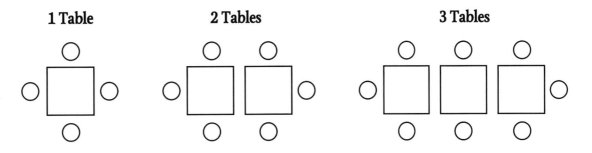

Work together to answer the following questions. When everyone agrees on an answer, write it in the space provided.

1. How many chairs are needed for 4 tables? for 5 tables? _____

2. Predict the number of chairs that are needed for 12 tables. _____

3. Explain how you made this prediction. _____

    _____

    _____

    _____

4. Describe a rule for the pattern. That is, if you are given the number of tables, describe a rule for finding the number of chairs.

    _____

    _____

    _____

Algebra

Goal: To provide opportunities for students to develop concepts and skills related to coordinate graphing

NCTM Standards, Grades 6–8

Algebra

Use mathematical models to represent and understand quantitative relationships: model and solve contextualized problems using various representations, such as graphs, tables, and equations.

Student Activities Overview and Answer Key

Station 1

Students work together to complete a table for the relationship $y = 3x$. They use the completed table to write ordered pairs that satisfy the equation and then work together to plot these points on a coordinate plane. Once students have plotted the points, they complete the graph and describe it.

Answers: 1. See below for table. 2. (0, 0), (1, 3), (2, 6), (3, 9), (4, 12); 3–5. See below for graph.
6. Possible answer: The graph is a straight line that passes through the origin.

x	y
0	0
1	3
2	6
3	9
4	12

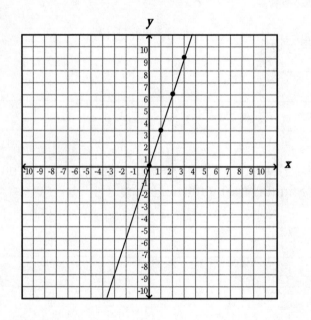

Station 2

In this activity, students are given a table of values. They work together to plot these points and look for patterns. (The points all satisfy the relationship $y = 1.5x$, so they all lie on a straight line.) Using the graph of plotted points, students name an additional point that they think satisfies the same relationship and they justify their choice.

Answers: 1. (–4, –6), (–2, –3), (1, 1.5), (4, 6), (5, 7.5); 2–3. See below for graph. 4. The points all lie on a straight line. 5. Possible answer: (–6, –9) or (2, 3) 6. The point lies along the line determined by the other points.

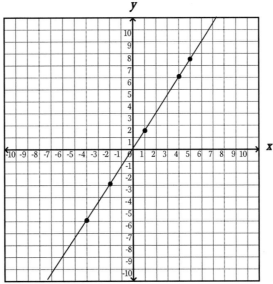

Station 3

Students work together to explore a pattern made from tiles. They complete a table showing the relationship between the stage of the pattern and the number of tiles needed. Then they plot the points (which lie on the line $y = 4x$) and use the graph to predict the number of tiles that would be needed to make Stage 6 of the pattern.

Answers: 1. See next page for table. 2. (1, 4), (2, 8), (3, 12), (4, 16), (5, 20); 3–4. See next page for graph. 5. 24 tiles; 6. Possible explanation: The plotted points all lie on a line. Continue the pattern of points to find that the next one is (6, 24).

Stage (x)	1	2	3	4	5
Number of Tiles (y)	4	8	12	16	20

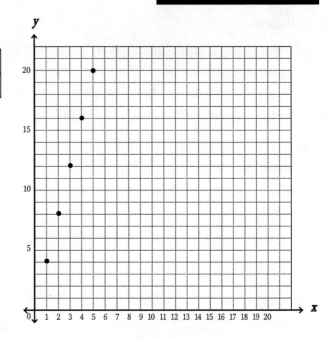

Station 4

Students work together to complete a table for the relationship $y = -2x$. They use the completed table to write ordered pairs that satisfy the equation and then work together to plot these points on a coordinate plane. Once students have plotted the points, they complete the graph and describe it.

Answers: 1. See below for table. 2. (–3, 6), (–1, 2), (0, 0), (2, –4), (4, –8); 3–5. See below for graph.
6. Possible answer: The graph is a downward-sloping straight line that passes through the origin.

x	y
–3	6
–1	2
0	0
2	–4
4	–8

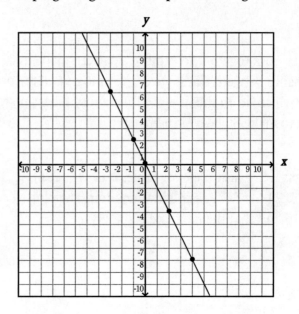

Materials List/Set Up

Station 1   graph paper; ruler

Station 2   graph paper; ruler

Station 3   small square tiles; graph paper; ruler

Station 4   graph paper; ruler

Discussion Guide

To support students in reflecting on the activities, and to gather formative information about student learning, use the following prompts to facilitate a class discussion to "debrief" the station activities.

Prompts/Questions

1. How do you set up a coordinate plane on a sheet of graph paper?

2. How do you plot an ordered pair on a coordinate plane?

3. How do you get ordered pairs from a table of values?

4. What can you say about the graph of any equation that has the form $y = kx$?

Think, Pair, Share

Have students jot down their own responses to questions, discuss their responses with a partner (who was not in their station group), and then discuss as a whole class.

Suggested Appropriate Responses

1. Use a ruler to draw a straight line (x-axis). Draw another line at a right angle to the first line (y-axis). Label the intersection as the origin, (0, 0).

2. First plot the x-value by moving along the x-axis by the given number of units. Then move up (or down) from this point to plot the y-value.

3. In each row or column, take the x-value and the corresponding y-value, in that order, to form one ordered pair.

4. The graph is a straight line that passes through the origin.

Possible Misunderstandings/Mistakes

- Plotting points incorrectly due to reversing the x- and y-values (e.g., plotting (3, 4) rather than (4, 3))

- Plotting points incorrectly due to mislabeling or incorrectly calibrating the scale on the x- or y-axis

- Incorrectly reading the ordered pairs from a table of values

Algebra
Set 2: Graphing Relationships

Station 1

At this station, you will work together to make a table and graph for the relationship $y = 3x$.

x	y
0	
1	
2	6
3	
4	

The equation $y = 3x$ states that the value of y is 3 times the value of x.

1. Work with other students to complete the table of values.

2. Write the ordered pairs of values (x, y) from your table.

_____

3. Set up a coordinate plane on a sheet of graph paper. Use a ruler to draw the x-axis and the y-axis.

4. Plot the ordered pairs from your table. Work together to make sure that all the points are plotted correctly.

5. Use the points you plotted to draw the complete graph of $y = 3x$.

6. Describe the graph of $y = 3x$.

_____

_____

_____

_____

Algebra
Set 2: Graphing Relationships

Station 2

At this station, you will plot points from a table and use your graph to make predictions.

x	y
−4	−6
−2	−3
1	1.5
4	6
5	7.5

The table to the right shows a set of values. Work with other students to explore how the x- and y- values are related.

1. Write the ordered pairs of values, (x, y), from the table.

   _____

2. Set up a coordinate plane on a sheet of graph paper. Use a ruler to draw the x-axis and the y-axis.

3. Plot the ordered pairs. Work together to make sure that all the points are plotted correctly.

4. Describe what you notice about the points.

   _____

   _____

   _____

5. Add a new point to the graph that you think has the same relationship between the x-value and y-value. What are the coordinates of the point? _____

6. Explain how you decided on this point.

   _____

   _____

   _____

Algebra
Set 2: Graphing Relationships

Station 3

At this station, you will use a table and graph to explore a pattern. You can also use tiles to build the pattern.

Marissa is using tiles to make a pattern. Here are the first three stages in her pattern.

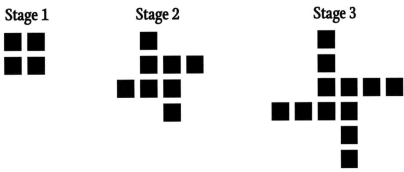

1. Complete the table.

Stage (x)	1	2	3	4	5
Number of Tiles (y)					

2. Write the ordered pairs of values (x, y) from the table.

_____

3. Set up a coordinate plane on a sheet of graph paper. Use a ruler to draw the x-axis and the y-axis.

4. Plot the ordered pairs. Work together to make sure that all the points are plotted correctly.

5. Use your graph to predict the number of tiles that are needed at Stage 6. _____

6. Explain how you made this prediction.

_____

_____

_____

Algebra
Set 2: Graphing Relationships

Station 4

At this station, you will work together to make a table and graph for the relationship $y = -2x$.

The equation $y = -2x$ states that the value of y is -2 times the value of x.

x	y
-3	
-1	
0	
2	-4
4	

1. Work with other students to complete the table of values.

2. Write the ordered pairs of values from your table.

3. Set up a coordinate plane on a sheet of graph paper. Use a ruler to draw the x-axis and the y-axis.

4. Plot the ordered pairs from your table. Work together to make sure that all the points are plotted correctly.

5. Use the points you plotted to draw the complete graph of $y = -2x$.

6. Describe the graph of $y = -2x$.

_____

_____

_____

_____

Algebra

Set 3: Evaluating Expressions

Goal: To provide opportunities for students to develop concepts and skills related to evaluating expressions

NCTM Standards, Grades 6–8

Algebra

Represent and analyze mathematical situations and structures using algebraic symbols: develop an initial conceptual understanding of different uses of variables.

Student Activities Overview and Answer Key

Station 1

In this activity, students work together to evaluate expressions. They choose an expression at random from a set of cards and then roll a number cube to get a value for the variable. Then they work together to evaluate the expression for this value of the variable. After repeating the process several times, students reflect on the strategies they used to evaluate the expressions.

Answers: Answers will depend upon the cards that are chosen and the numbers that are rolled.

Possible strategies: Substitute the value of the variable in the expression. Simplify the numerical expression using the order of operations.

Station 2

Students work together to evaluate the expression $2x + 1$ for a variety of values of x. They also choose their own values of x to substitute into the expression. After doing so, students look back on their results and look for a pattern. They find that no matter what counting number is used as the value of x, the result of evaluating $2x + 1$ is an odd number.

Answers: For $x = 1$, $2x + 1 = 3$; for $x = 2$, $2x + 1 = 5$; for $x = 3$, $2x + 1 = 7$; for $x = 7$, $2x + 1 = 15$; for $x = 9$, $2x + 1 = 19$. Other values of the expression will depend upon values chosen for the variable. No matter what counting number is used for the value of x, the value of $2x + 1$ is always odd.

Station 3

Students work together to complete tables by evaluating expressions for given values of the variable. Then students describe any patterns they see in the tables. In each case, the values of the variable increase by 1, while the values of the expression increase by a constant amount.

Answers:

Value of x	1	2	3	4	5
Value of $3x + 5$	8	11	14	17	20

Possible patterns: The values of x increase by 1; the values of $3x + 5$ increase by 3.

Value of x	3	4	5	6	7
Value of $10x - 2$	28	38	48	58	68

Possible patterns: The values of x increase by 1; the values of $10x - 2$ increase by 10.

Station 4

Students work together to match a set of given expressions with a set of values of the variable. The goal is to pair expressions and values of the variable so that every expression has a value of 12 when evaluated for the value of the variable with which it is paired. All students should agree on the pairing of the cards before writing the answer.

Answers: The cards should be paired as follows: $2x + 2$ and $x = 5$; $5 + x$ and $x = 7$; $36 \div x$ and $x = 3$; $16 - x$ and $x = 4$; $3x^2$ and $x = 2$.

Possible strategies: Choose an expression and evaluate it for each possible value of the variable until the result is 12. Alternatively, choose an expression and decide which value of the variable makes it equal 12, and then match it with this value.

Materials List/Set Up

Station 1	number cube (numbers 1–6)
	set of 5 index cards with the following expressions written on them:
	$3n + 4$, $2m^2$, $15 - x$, $60 \div p$, $2t - 2$
Station 2	none
Station 3	none
Station 4	set of 5 index cards with the following expressions written on them:
	$2x + 2$, $5 + x$, $36 \div x$, $16 - x$, $2m^2$
	set of 5 index cards with the following values of x written on them:
	$x = 2$, $x = 3$, $x = 4$, $x = 5$, $x = 7$

Discussion Guide

To support students in reflecting on the activities, and to gather formative information about student learning, use the following prompts to facilitate a class discussion to "debrief" the station activities.

Prompts/Questions

1. How do you evaluate an expression for a given value of the variable?

2. After you substitute the value of the variable in the expression, how do you simplify the result?

3. How do you evaluate the expression $2n^2$ for a specific value of n?

4. What strategies can you use to help you find patterns in a table of values?

Think, Pair, Share

Have students jot down their own responses to questions, discuss their responses with a partner (who was not in their station group), and then discuss as a whole class.

Suggested Appropriate Responses

1. Substitute the value of the variable for the variable in the expression and simplify.

2. Perform the operation(s), using the order of operations if necessary.

3. Substitute the value for n. Then square the value by multiplying it by itself. Then multiply by 2.

4. Look for patterns as you go from one value to the next. For example, check to see if the values increase or decrease by a fixed number each time.

Possible Misunderstandings/Mistakes

- Applying an incorrect operation (e.g., adding instead of multiplying when evaluating for a value of x)

- Evaluating exponents incorrectly

- Incorrectly applying the order of operations when evaluating an expression that involves more than one operation

Algebra
Set 3: Evaluating Expressions

Station 1

At this station, you will find a stack of cards and a number cube. Each card contains an algebraic expression.

Choose one of the cards without looking. Write the expression on the line. _____

Roll the number cube to get a value for the variable. Write this value on the line.

_____

Work with other students to evaluate the expression for this value of the variable. When everyone agrees on the result, write it below.

_____

Put the card back. Mix up the cards. Repeat the above process four more times.

Expression on card: _____

Value of variable: _____

Evaluate the expression: _____

Expression on card: _____

Value of variable: _____

Evaluate the expression: _____

Expression on card: _____

Value of variable: _____

Evaluate the expression: _____

Expression on card: _____

Value of variable: _____

Evaluate the expression: _____

Describe the strategies you used to evaluate the expressions for the given values of the variable.

_____

_____

_____

Algebra
Set 3: Evaluating Expressions

Station 2

At this station, you will work with other students to evaluate expressions and look for patterns.

Work together to evaluate the expression $2x + 1$ for each value of the variable shown below. When everyone agrees on an answer, write it down.

$x = 1$ _____

$x = 2$ _____

$x = 3$ _____

$x = 7$ _____

$x = 9$ _____

Choose your own three values for x. (The values should be counting numbers, such as 1, 2, 3, and so on.) Then evaluate the expression $2x + 1$ for these values of x.

Value of x: _____

Evaluate the expression: _____

Value of x: _____

Evaluate the expression: _____

Value of x: _____

Evaluate the expression: _____

No matter what counting number you substitute for x, what is true about the result when you evaluate $2x + 1$?

_____

_____

_____

Algebra
Set 3: Evaluating Expressions

Station 3

At this station, you will work with other students to complete tables and look for patterns.

Work with other students to complete this table. To do so, evaluate $3x + 5$ for each value of x in the table.

Value of x	1	2	3	4	5
Value of $3x + 5$					

Look for patterns in the table. Describe at least two patterns that you see.

_____

_____

_____

_____

Work together to complete this table. This time, evaluate $10x - 2$ for each value of x in the table.

Value of x	3	4	5	6	7
Value of $10x - 2$					

Look for patterns in the table. Describe at least two patterns that you see.

_____

_____

_____

_____

Algebra
Set 3: Evaluating Expressions

Station 4

At this station, you will find five cards with these expressions written on them.

$$2x + 2 \qquad 5 + x \qquad 36 \div x \qquad 16 - x \qquad 3x^2$$

You will also find five cards with these values of x written on them.

$$x = 2 \qquad x = 3 \qquad x = 4 \qquad x = 5 \qquad x = 7$$

Work together to match each expression with a value of x. When you evaluate each expression for its value of x, the result should be 12.

Work together to check that each pair gives a value of 12.

When everyone agrees on the results, write the five pairs below.

_____

_____

_____

_____

_____

Describe the strategies you used to solve this problem.

_____

_____

_____

_____

Algebra

Goal: To provide opportunities for students to develop concepts and skills related to solving equations

NCTM Standards, Grades 6–8

Algebra

Represent and analyze mathematical situations and structures using algebraic symbols: develop an initial conceptual understanding of different uses of variables.

Student Activities Overview and Answer Key

Station 1

Students are given a set of cards with simple one-step equations written on them. They are given another set of cards with values of the variable written on them. Students work together to match each equation to its solution. Once students have paired the cards, they reflect on the strategies they used.

Answers: The cards should be paired as follows: $8 + x = 12$ and $x = 4$; $x/4 = 2$ and $x = 8$; $6x = 18$ and $x = 3$; $x - 4 = 2$ and $x = 6$; $14 = 7x$ and $x = 2$; $2 = x/6$ and $x = 12$.

Possible strategies: Choose an equation. Check each value of x in the equation to see if it is a solution. Alternatively, solve the equation and look for its solution among the values of x.

Station 2

Students work together to use algebra tiles to represent simple one-step equations. Then they use the tiles to help them solve the equations. Students explain the strategies they used to manipulate the tiles and solve the equations.

Answers: 1. $x = 3$; 2. $x = 6$; 3. $x = 2$; 4. $x = 2$; 5. $x = 6$; 6. $x = 3$

Possible strategies: Remove the same number of yellow tiles from both sides of the mat. Divide the tiles on each side of the mat into the same number of equal groups, and remove all but one of the groups on each side. The remaining group shows the solution.

Station 3

In this activity, students use cups and counters to model simple one-step equations. In the given pictures, each cup is holding an unknown number of counters. Students use this idea to write the equation that is modeled by each picture. Then they use actual cups and counters, as well as logical reasoning, to help them find the unknown number of counters in each cup. This is equivalent to solving the corresponding equation.

Answers: 1. $x + 2 = 9$, $x = 7$; 2. $6 = x + 1$, $x = 5$; 3. $2x = 10$, $x = 5$; 4. $3x = 12$, $x = 4$

Station 4

Students are given a set of equations and a set of real-world situations. They work together to match each situation to an equation. Then they solve the equation. At the end of the activity, students explain the strategies they used to match the equations to the situations.

Answers: 1. $x + 6 = 18$, $x = 12$; 2. $x/6 = 18$, $x = 108$; 3. $6x = 18$, $x = 3$; 4. $x - 6 = 18$, $x = 24$

Possible strategies: Use the words or phrases that refer to arithmetic operations as clues to identifying the corresponding equations. For example, separating into equal groups corresponds to division.

Materials List/Set Up

Station 1   set of 6 index cards with the following equations written on them:
$8 + x = 12$, $x/4 = 2$, $6x = 18$, $x - 4 = 2$, $14 = 7x$, $2 = x/6$
set of 6 index cards with the following values of x written on them:
$x = 2$, $x = 3$, $x = 4$, $x = 6$, $x = 8$, $x = 12$

Station 2   algebra tiles and equation mat

Station 3   3 paper cups; 12 counters or other small objects, such as pennies or beans

Station 4   none

Discussion Guide

To support students in reflecting on the activities, and to gather formative information about student learning, use the following prompts to facilitate a class discussion to "debrief" the station activities.

Prompts/Questions

1. What are some different tools, objects, or drawings that you can use to help you solve equations?

2. What does the equal sign tell you in an equation?

3. How do you solve an equation using inverse operations?

4. How can you check your solution to an equation?

Think, Pair, Share

Have students jot down their own responses to questions, discuss their responses with a partner (who was not in their station group), and then discuss as a whole class.

Suggested Appropriate Responses

1. algebra tiles, cups and counters, drawings of balance scales, etc.

2. It tells you that the quantities on either side of the equation are the same.

3. Isolate the variable by applying inverse operations to both sides of the equation.

4. Substitute the value for the variable in the equation and simplify. If the solution is correct, the two sides of the equation should be equal.

Possible Misunderstandings/Mistakes

- Using an incorrect operation to solve an equation (e.g., solving $x + 3 = 12$ by adding 3 to both sides)

- Attempting to solve an equation such as $x + 4 = 9$ by subtracting x from both sides

- Applying an operation that does not isolate the variable (e.g., solving $9 = 3x$ by dividing both sides by 9)

Algebra
Set 4: Solving Equations

Station 1

At this station, you will find a set of cards with the following equations written on them:

$$8 + x = 12 \qquad \frac{x}{4} = 2 \qquad 6x = 18 \qquad x - 4 = 2 \qquad 14 = 7x \qquad 2 = \frac{x}{6}$$

You will also find a set of cards with the following values of x written on them:

$$x = 2 \qquad x = 3 \qquad x = 4 \qquad x = 6 \qquad x = 8 \qquad x = 12$$

Work with other students to match each equation with its solution.

Work together to check that each equation is paired with its correct solution. Write the pairs below.

_____

_____

_____

_____

_____

_____

Explain the strategies you used to match up the cards.

_____

_____

_____

_____

Algebra
Set 4: Solving Equations

Station 2

You can use algebra tiles to help you solve equations.

Each square yellow tile shows +1. Each square red tile shows −1. Each rectangular yellow tile shows x. You use the equation mat to show the two sides of an equation.

Work together to use algebra tiles to show each equation. Then use the tiles to solve each equation. Write the value of x for each equation below.

1. $x + 5 = 8$ _____

2. $9 = x + 3$ _____

3. $4 + x = 6$ _____

4. $3x = 6$ _____

5. $2x = 12$ _____

6. $9 = 3x$ _____

Explain at least two strategies you used to solve the equations using algebra tiles.

_____

_____

_____

_____

Algebra
Set 4: Solving Equations

Station 3

In each picture, the cup is holding an unknown number of counters, x. If there is more than one cup, every cup is holding the same number of counters.

Each picture shows an equation. This picture shows $x + 5 = 7$. To make the two sides equal, there must be 2 counters in the cup. This means $x = 2$.

Work with other students to write an equation for each picture. Then find the number of counters in each cup. You can use the cups and counters at the station to help you.

1. Equation: _____

   Solution: _____

2. Equation: _____

   Solution: _____

3. Equation: _____

   Solution: _____

4. Equation: _____

   Solution: _____

Algebra
Set 4: Solving Equations

Station 4

At this station, you will match equations to real-world situations and then solve the equations.

Work with other students to match each situation to one of the following equations. When everyone agrees on the correct equation, write it in the space provided. Then work together to solve it.

$$\frac{x}{6} = 18 \qquad x - 6 = 18 \qquad 6x = 18 \qquad x + 6 = 18$$

1. In 6 years, Rosario will be 18 years old. How old is she now?

   Equation: _____

   Solution: _____

2. When the students at Essex Middle School are separated into 6 equal groups, there are 18 students in each group. How many students are at the school?

   Equation: _____

   Solution: _____

3. Mike bought some books that cost $6 each. He spent a total of $18. How many books did he buy?

   Equation: _____

   Solution: _____

4. From 4:00 to 5:00 P.M., the temperature dropped 6°F. At 5:00 P.M. the temperature was 18°F. What was the temperature at 4:00 P.M.?

   Equation: _____

   Solution: _____

Explain the strategies you used to match the equations to the situations.

_____

_____

_____

Algebra

Set 5: Proportional Relationships

Goal: To provide opportunities for students to develop concepts and skills related to proportional relationships

NCTM Standards, Grades 6–8

Algebra

Understand patterns, relations, and functions: relate and compare different forms of representation for a relationship.

Use mathematical models to represent and understand quantitative relationships: model and solve contextualized problems using various representations, such as graphs, tables, and equations.

Student Activities Overview and Answer Key

Station 1

Students are given a set of cards with ratios written on them. Students work together to pair up the cards so that the ratios shown on each pair of cards are equal. After forming the pairs of cards, students explain the strategies they used to solve the problem.

Answers: The cards should be paired as follows: $\frac{2}{3} = \frac{6}{9}$; $\frac{4}{8} = \frac{8}{16}$; $\frac{5}{15} = \frac{2}{6}$; $\frac{12}{15} = \frac{8}{10}$; $\frac{3}{4} = \frac{6}{8}$

Possible strategies: Look for pairs of cards that represent common ratios, such as 1:2. Reduce fractions to lowest terms and look for equivalent pairs. Pair up cards and check to see if the ratios are equal using cross products.

Station 2

Students work together to explore a real-world situation that involves a proportional relationship. First, students look for patterns in a table of data and complete the table. Then they make predictions about the data in the table and write an equation in the form of $y = kx$ that describes the relationship.

Answers: 1. See below for complete table. 2. Possible explanation: The number of ounces of meat divided by the number of feedings is always 2.4. 3. 2.4 ounces; 4. 48 ounces; 5. $y = 2.4x$; 6. Since x is the number of feedings, you must multiply it by 2.4 to get the number of ounces of meat. So the relationship is $y = 2.4x$.

Number of Feedings	3	4	6	9	12
Ounces of Meat	7.2	9.6	14.4	21.6	28.8

Station 3

Students learn a step-by-step method for using graph paper and a straightedge to check whether two ratios are proportional. Then students work together to use this method to test several pairs of ratios. They also use the method to find a ratio that is proportional to a given ratio.

Answers: 1. yes; 2. yes; 3. no; 4. yes; 5. no; 6. yes; 7. Possible ratio: $1/4$; Draw a 3-by-12 rectangle and its diagonal. Then look for a smaller or larger rectangle that could have the same line as a diagonal.

Station 4

Students work together to use the numbers 2, 3, 4, and 6 to write as many different proportions as they can. To help them, the numbers are provided on cards that can be moved around and paired in different ways. Students explain the strategies they used to find the proportions.

Answers: Possible proportions: $2/3 = 4/6$; $2/4 = 3/6$; $3/2 = 6/4$; $4/2 = 6/3$

Possible strategies: Place the numbers in any positions and use cross products to check whether the ratios form a proportion. Try to pair the numbers to make common ratios, such as 1:2.

Materials List/Set Up

Station 1	set of 12 index cards with the following ratios written on them: $2/3$, $6/9$, $8/10$, $12/15$, $3/4$, $6/8$, $4/8$, $8/16$, $2/6$, $5/15$
Station 2	calculator
Station 3	graph paper; straightedge
Station 4	set of 4 index cards with the following numbers written on them: 2, 3, 4, 6

Discussion Guide

To support students in reflecting on the activities, and to gather formative information about student learning, use the following prompts to facilitate a class discussion to "debrief" the station activities.

Prompts/Questions

1. Is the ratio $^2/_3$ the same as the ratio $^3/_2$? Why or why not?

2. What does it mean when we say that two ratios form a proportion?

3. How can you check that two ratios form a proportion?

4. If you are given a ratio, how can you form a different ratio that is proportional to the given one?

Think, Pair, Share

Have students jot down their own responses to questions, discuss their responses with a partner (who was not in their station group), and then discuss as a whole class.

Suggested Appropriate Responses

1. No. The ratio $^2/_3$ is less than 1, while the ratio $^3/_2$ is greater than 1.

2. The ratios are equivalent. They name the same fraction.

3. Check to see if the cross products are equal.

4. Form a new ratio by multiplying the numerator and denominator of the given ratio by the same number.

Possible Misunderstandings/Mistakes

- Assuming that a ratio and its reciprocal are equivalent ratios
- Adding a number to the numerator and denominator of a given ratio to form a new ratio that is proportional to the given one
- Failing to recognize that whole numbers may be written as ratios by using a denominator of 1

Algebra
Set 5: Proportional Relationships

Station 1

At this station, you will find a set of cards with the following ratios written on them:

$$\frac{2}{3} \qquad \frac{4}{8} \qquad \frac{5}{15} \qquad \frac{12}{15} \qquad \frac{3}{4} \qquad \frac{6}{8} \qquad \frac{6}{9} \qquad \frac{8}{16} \qquad \frac{2}{6} \qquad \frac{8}{10}$$

Work with other students to put the cards in pairs. The ratios in each pair should be proportional.

Work together to check that the ratios are proportional. When everyone agrees on your answers, write the pairs of equal ratios below.

_____

_____

_____

_____

_____

Explain at least three strategies that you used to solve this problem.

_____

_____

_____

_____

_____

_____

_____

Algebra
Set 5: Proportional Relationships

Station 2

At this station, you will work with other students to explore a real-world situation.

A zookeeper is taking care of a baby tiger. The table below shows the number of feedings and the number of ounces of meat that are needed.

Number of Feedings	3	4	6	9	12
Ounces of Meat	7.2	9.6	14.4		

Work with other students to look for patterns in the table. You may use a calculator to help you.

1. Complete the table.

2. Explain how you completed the table. _____

   _____

   _____

   _____

3. How many ounces of meat are needed for each feeding? _____

4. How many ounces of meat are needed for 20 feedings? _____

5. Write an equation that shows the number of ounces of meat y that are needed for x feedings.

   _____

6. Explain how you came up with the equation.

   _____

   _____

   _____

Algebra
Set 5: Proportional Relationships

Station 3

You will find graph paper and a straightedge at this station. You will use these tools to check whether ratios are proportional.

To check whether $\dfrac{2}{3}$ and $\dfrac{4}{6}$ are proportional, first draw a 2-by-3 rectangle on the graph paper.

Now draw a 4-by-6 rectangle. The two rectangles should share a corner as shown.

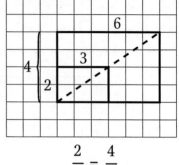

$$\frac{2}{3} = \frac{4}{6}$$

Use the straightedge to draw the diagonals of the rectangles. If you can draw the diagonals of both rectangles with one straight line, then the ratios are proportional.

Work with other students to use this method to check whether the ratios in each pair are proportional. Write "yes" or "no" for each pair.

1. $\dfrac{1}{2}$ and $\dfrac{4}{8}$ _____

2. $\dfrac{6}{8}$ and $\dfrac{9}{12}$ _____

3. $\dfrac{5}{8}$ and $\dfrac{10}{14}$ _____

4. $\dfrac{2}{5}$ and $\dfrac{4}{10}$ _____

5. $\dfrac{6}{10}$ and $\dfrac{8}{12}$ _____

6. $\dfrac{4}{6}$ and $\dfrac{6}{9}$ _____

7. Use this method to find a ratio that is proportional to $\dfrac{3}{12}$. Write the ratio below and explain your work.

_____

_____

_____

_____

Algebra
Set 5: Proportional Relationships

Station 4

At this station, you will work with other students to form ratios that are proportional.

You will find cards at this station with the numbers 2, 3, 4, and 6 on them.

Work with other students to write the four numbers in the boxes below so that the two ratios are proportional.

$$\frac{\Box}{\Box} = \frac{\Box}{\Box}$$

Use the cards to help you move the numbers around and form other ratios that are proportional. Work together to find as many different proportions as you can.

Record your proportions below.

_____

_____

_____

Explain the strategies you used to find the proportions.

Data Analysis and Probability

Set 1: Collecting, Organizing, and Analyzing Data

Goal: To provide opportunities for students to develop concepts and skills related to data collection and analysis

NCTM Standards, Grades 6–8

Data Analysis and Probability

Formulate questions that can be addressed with data and collect, organize, and display relevant data to answer them: formulate questions, design studies, and collect data about a characteristic shared by two populations or different characteristics within one population; select, create, and use appropriate graphical representations of data, including histograms, box plots, and scatterplots.

Student Activities Overview and Answer Key

Station 1

Students will survey their classmates and turn that data into a stem-and-leaf plot. They will use the plot to analyze the data, and then reflect on how a stem-and-leaf plot is helpful.

Answers: Answers will vary; answers will vary; a stem-and-leaf plot helps us organize our data so it is easier to read.

Station 2

Students will be generating their own data and turning it into a scatter plot. They will examine and answer questions pertaining to this scatter plot.

Answers: Answers will vary.

Station 3

Students will formulate different questions for which they would be able to go out and collect data. They then choose one question to look at more closely. Students reflect on how they would work toward answering that question.

Answers: Answers will vary.

Station 4

Students try to answer the question of which comes up more, heads or tails, when flipping a coin. They conduct an experiment to help answer this question, then extend that to 10,000 flips.

Answers: Answers will vary; heads and tails should come up about the same number of times.

Materials List/Set Up

Station 1   none

Station 2   ruler; two number cubes (1–6)

Station 3   none

Station 4   pennies for all group members

Discussion Guide

To support students in reflecting on the activities, and to gather formative information about student learning, use the following prompts to facilitate a class discussion to "debrief" the station activities.

Prompts/Questions

1. How does conducting an experiment help us answer questions?

2. What types of questions are best answered by numerical data?

3. What is an example of a real-life situation where we might want to conduct a survey to answer a question?

4. When would it be a good idea to use a stem-and-leaf plot in a real-world situation?

Think, Pair, Share

Have students jot down their own responses to questions, discuss their responses with a partner (who was not in their station group), and then discuss as a whole class.

Suggested Appropriate Responses

1. It lets us see how frequently we get different outcomes and what those outcomes are.

2. questions that ask how many

3. Anything where we care what a specific population does, wants, etc. For example, how much money do most people spend at lunch?

4. when you want to organize data numerically

Possible Misunderstandings/Mistakes

- Not putting data in order for the stem-and-leaf plot
- Not appropriately labeling the bar graph
- Asking questions that are too broad

Data Analysis and Probability
Set 1: Collecting, Organizing, and Analyzing Data

Station 1

You will be surveying the students in your class. Choose one of the following questions to use for your research.

- How many hours per week do you spend on homework?
- How many hours per week do you spend watching TV?
- How many hours per week do you spend listening to music?

All class members should answer the question that your group chose. Keep track of the answers in a list on the lines below.

_____

_____

_____

Use this data to construct a stem-and-leaf plot.

Which piece of data is the median number of the set? _____

Which number of hours is the mode? _____

How does a stem-and-leaf plot help you analyze data? _____

_____

_____

Data Analysis and Probability
Set 1: Collecting, Organizing, and Analyzing Data

Station 2

At this station, you will find two number cubes. You will be using these number cubes to generate data which you will then graph.

Roll the number cubes one at a time. Record the number that comes up on the first number cube. Roll the second number cube and record the number. Then record the sum of the two number cubes. Do this 10 times and record the data in the table below.

First Number Cube	Second Number Cube	Sum of Number Cubes

Using this data, create a scatter plot in the space provided. Be sure to title the scatter plot and use appropriate labels.

What is the range of the sums? _____

What do you notice about your scatter plot? Is there a general trend? _____

_____

_____

Data Analysis and Probability
Set 1: Collecting, Organizing, and Analyzing Data

Station 3

At this station, you will come up with different questions you could answer by collecting data.

Work with your group to come up with five questions you could answer by going out and collecting data (e.g., What is the average number of siblings students in our class have?).

_____

_____

_____

_____

_____

Choose one of those questions to focus on.

How would you gather data to answer your question? _____

_____

_____

What materials would you need to gather the data (e.g., number cubes, etc.)? _____

_____

_____

How would you present your data (e.g., bar graph, table, etc.)?

_____

_____

Data Analysis and Probability
Set 1: Collecting, Organizing, and Analyzing Data

Station 4

At this station, you will try to answer the question of which side of a penny comes up more, heads or tails.

Each member of your group should flip a penny 25 times, keeping track of how many times heads and tails come up.

	Tally	Final Number
Heads		
Tails		

Combine your group's data in the table below.

	Total
Heads	
Tails	

Based on this data, which do you think has a better chance of coming up, heads or tails? Why?

_____

_____

_____

What do you think would happen if you flipped a penny 10,000 times? Do you think either heads or tails would come up more than the other? Why?

_____

_____

_____

Data Analysis and Probability

Set 2: Constructing Frequency Distributions

Goal: To provide opportunities for students to develop concepts and skills related to frequency distributions, frequency tables, and graphs

NCTM Standards, Grades 6–8

Data Analysis and Probability

Formulate questions that can be addressed with data and collect, organize, and display relevant data to answer them: select, create, and use appropriate graphical representations of data, including histograms, box plots, and scatterplots.

Student Activities Overview and Answer Key

Station 1

Students will look at a newspaper article to determine letter frequency. They will count the number of each letter in an article. They will then use this information to construct a frequency table.

Answers: Table: answers will vary; answers will vary: yes or no; answers will vary: probably e or t; answers will vary: probably q, x, or z; The tally column makes it easier to add up all the times each letter appears in the article.

Station 2

Students will look at the average amount of monthly rainfall for New York City. They construct a line graph using this information, and then answer questions based on their line graph.

Answers: May and November; 1.1 inches; You can easily see the general change in average rainfall amounts by looking at the graph.

Station 3

Students roll a number cube and put their data into a frequency table. They use this data to answer appropriate questions and reflect on how the distribution table helped them.

Answers: Answers will vary; The Tally column means that you don't have to write down all the numbers you roll. Instead, you just put a mark, which means you rolled a given number.

Station 4

Students will work as a group to analyze a scatter plot. They will draw conclusions based on their observations, and use those conclusions to make general statements and predictions.

Answers: Number of hours spent studying and test scores; The more a student studies for the test, the higher that student's score; between a 65 and 70; We looked at where one half is on the scatter plot, and looked at the other points around that area.

Materials List/Set Up

Station 1	newspaper article—150 words or less in length
Station 2	ruler
Station 3	number cube (1–6)
Station 4	none

Discussion Guide

To support students in reflecting on the activities, and to gather formative information about student learning, use the following prompts to facilitate a class discussion to "debrief" the station activities.

Prompts/Questions

1. How do frequency tables help us sort data?

2. What conclusions can you draw from a scatter plot if all the points are in a line? if the points are randomly distributed?

3. What is a good, real-life situation to model with a scatter plot?

4. When should a line graph be used?

Think, Pair, Share

Have students jot down their own responses to questions, discuss their responses with a partner (who was not in their station group), and then discuss as a whole class.

Suggested Appropriate Responses

1. We use the tally column to keep track of our data as we collect it, which is an organized strategy. They are also in numerical order, which makes it easier to read.

2. If the points are in a line, there is a correlation between the two variables. If the points are randomly distributed, one variable does not affect the other.

3. anything that involves two variables that have a relationship—one affects the other

4. when we are looking at a variable over time

Possible Misunderstandings/Mistakes

- Losing track of which letters have been counted
- Mislabeling the graph/axes
- Having trouble creating a scale for numbers with decimal endings

NAME:

Data Analysis and Probability
Set 2: Constructing Frequency Distributions

Station 1

At this station, you will be learning about the frequency that letters appear in a newspaper article. With your group, count the number of each letter in the article. Record your data in the table below. Place a mark for each letter in the column titled "Tally." Then write the total number of marks in the "Total Number" column after you have finished counting all the letters.

Letter	Tally	Total Number	Letter	Tally	Total Number
A			N		
B			O		
C			P		
D			Q		
E			R		
F			S		
G			T		
H			U		
I			V		
J			W		
K			X		
L			Y		
M			Z		

E is the most frequently used letter in the English language. Do your results support this statement?

_____

What letter(s) appears the most frequently? _____

What letter(s) appears the least frequently? _____

How does the "Tally" column help you organize your data? _____

_____

Data Analysis and Probability
Set 2: Constructing Frequency Distributions

Station 2

At this station, you will use a table to answer questions.

The average rainfall in New York City for each month is shown in the table below.

Month	Jan	Feb	Mar	Apr	May	June	July	Aug	Sept	Oct	Nov	Dec
Amount of Rain (in.)	3.3	3.1	3.9	3.7	4.2	3.3	4.1	4.1	3.6	3.3	4.2	3.6

Source: http://www.worldclimate.com/cgi-bin/data.pl?ref=N40W073+2200+305801C

Construct a line graph using this information in the space below.

What month has the most rainfall on average? _____

What is the range of rainfall? _____

Why is a line graph an appropriate way to display this data? _____

_____

_____

Data Analysis and Probability
Set 2: Constructing Frequency Distributions

Station 3

At this station, you will find a number cube.

Roll the number cube 30 times. Record your data in the distribution table below.

Number	Tally	Total Occurrences
1		
2		
3		
4		
5		
6		

What do you notice about the overall trend of the data? Did you roll about the same amount of each number?

_____

_____

_____

How does the "Tally" column help you organize your data?

_____

_____

_____

Data Analysis and Probability
Set 2: Constructing Frequency Distributions

Station 4

Discuss the following scatter plot with your group members. Then work together to answer the questions that follow.

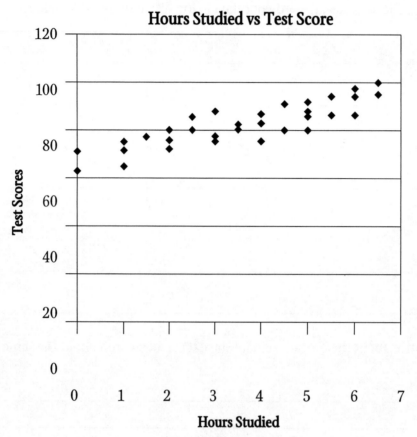

What two pieces of data are being compared in this scatter plot? _____

_____

What do you notice about the general trend of the data in this scatter plot? _____

_____

If you were told that a student spent half an hour studying, approximately what grade would you expect that student to earn on the test?

_____

What information did you use to make your last prediction? _____

_____

Data Analysis and Probability

Set 3: Using Tables and Graphs

Goal: To provide opportunities for students to develop concepts and skills related to using tables and graphs to examine variation in data

NCTM Standards, Grades 6–8

Data Analysis and Probability

Select and use appropriate statistical methods to analyze data: discuss and understand the correspondence between data sets and their graphical representations, especially histograms, stem-and-leaf plots, box plots, and scatterplots.

Student Activities Overview and Answer Key

Station 1

Students review the average monthly highs and lows for Atlanta, GA. They use this information to construct a double line graph. They then answer questions pertaining to their graph and reflect on why it is useful.

Answers: No, the lows will never be higher than the highs; the range; it allows us to see the differences between the highs and the lows and also see how the average temperature varies from month to month.

Station 2

Students will be generating their own data and turning it into a scatter plot. They will examine and answer questions pertaining to this scatter plot.

Answers: Answers will vary.

Station 3

Students will be looking at the make-up of packs of fun-size Skittles®. They will create a table using the Skittles, and determine the variation in their data. Ultimately, they will draw conclusions about the make-up of a random pack of Skittles.

Answers: Answers will vary; possibly that approximately the same number of each color is in each bag

Station 4

Students review the scores from Akron Acorns' games in February 2007. They then calculate the mean and range for the data, and answer questions based on these numbers. They look at the variation between the two teams.

Answers: 2.6; 3.8; 5 goals; 4 goals; It looks like the Acorns lost more because their mean score is lower and they had more variation in their data. The opponents were consistently higher.

Materials List/Set Up

Station 1	ruler
Station 2	ruler, 2 number cubes (1–6)
Station 3	fun-size packs of Skittles® (original fruit flavor) for every group member
Station 4	none

Discussion Guide

To support students in reflecting on the activities, and to gather formative information about student learning, use the following prompts to facilitate a class discussion to "debrief" the station activities.

Prompts/Questions

1. What is a real-life situation when you might want to know the range of data?

2. When is a double line graph a better choice than a scatter plot?

3. If the range in a set of data is very large, what does that say about the data?

4. What are some other ways to compare variation between two sets of data, besides looking at their means and range?

Think, Pair, Share

Have students jot down their own responses to questions, discuss their responses with a partner (who was not in their station group), and then discuss as a whole class.

Suggested Appropriate Responses

1. Possible answer: the cost of CDs at a store

2. when comparing two different values over time

3. There is a lot of variation; not consistent

4. looking at the medians

Possible Misunderstandings/Mistakes

- Not appropriately labeling the axes on graphs

- Having difficulty plotting two lines on the same graph

- Counting both end numbers in the range (e.g., smallest number 2, greatest number 6, and believing that since 2, 3, 4, 5, 6 is 5 numbers, the range is 5)

Data Analysis and Probability
Set 3: Using Tables and Graphs

Station 1

At this station, you will be constructing a double line graph.

Below are the average monthly highs and lows for Atlanta, GA.

Month	Low	High
January	33	52
February	36	57
March	44	65
April	50	73
May	60	80
June	67	87
July	71	89
August	70	88
September	64	82
October	53	73
November	44	63
December	36	55

Source: http://weather.msn.com/monthly_averages.aspx?wealocations=wc:USGA0028

Use this data to construct a double line graph in the space below. Be sure to title the graph and use appropriate labels. Then answer the questions on the next page.

continued

Data Analysis and Probability
Set 3: Using Tables and Graphs

Station 1, *continued*

Do the two lines ever cross? Why or why not? _____

_____

_____

What does the distance between the two lines represent? _____

_____

_____

Why is a double line graph an appropriate way to display this data?_____

_____

_____

Data Analysis and Probability
Set 3: Using Tables and Graphs

Station 2

At this station, you will find two number cubes. You will be using these number cubes to generate data which you will then graph.

Roll the number cubes one at a time. Record the number that comes up on the first number cube. Roll the second number cube, and record the number. Then record the sum of the two number cubes. Do this 10 times, and record the data in the table below.

First Number Cube	Second Number Cube	Sum of Number Cubes

Using this data, create a scatter plot in the space below. Be sure to title the scatter plot and use appropriate labels.

What is the range of the sums? _____

What do you notice about your scatter plot? Is there a general trend? _____

_____

Data Analysis and Probability
Set 3: Using Tables and Graphs

Station 3

At this station, each group member will receive a fun-size bag of Skittles®. You may not eat these because you need to use them to find how much the contents vary between bags.

Each group member should open his/her bag of Skittles. Record the number of each color of Skittle in the table below.

Group Member	Purple	Red	Yellow	Orange	Green
Total:					

What is the mean number of red Skittles in each bag? _____

What is the overall mean number of Skittles in each bag? _____

What is the range number of green Skittles? _____

What is the range number when looking at the total for each color? _____

Are the totals about the same for each color? _____

What conclusions can you draw about the make-up of the average bag of fun-size Skittles?

_____

_____

Data Analysis and Probability

Set 3: Using Tables and Graphs

Station 4

At this station, you will be looking at the scores from the Akron Acorns' games in February 2007. Below is a table with the results.

Akron Acorns' Score	Opponents' Score
2	5
2	5
3	4
6	3
2	3
1	5
1	4
3	5
3	1
4	5
1	4
3	2

What is the mean score for the Acorns? _____

What is the mean score for the opponents? _____

What is the range of scores for the Acorns? _____

What is the range of scores for the opponents? _____

Based only on your work above, do you think the opponents won or lost more games in February 2007? Why?

_____

_____

_____

Data Analysis and Probability

Set 4: Theoretical Probability

Goal: To provide opportunities for students to develop concepts and skills related to theoretical probability

NCTM Standards, Grades 6–8

Data Analysis and Probability

Understand and apply basic concepts of probability: use proportionality and a basic understanding of probability to make and test conjectures about the results of experiments and simulations; compute probabilities for simple compound events, using such methods as organized lists, tree diagrams, and area models.

Student Activities Overview and Answer Key

Station 1

Students will look at two pairs of number cubes. They will compare the probability of rolling particular sums. In the end, they will draw conclusions about the two pairs of number cubes.

Answers: $1/6$; $5/36$; $1/6$; $5/36$; The probability of rolling each sum is the same for both pairs of number cubes.

Station 2

Students will consider two different spinners. They determine the theoretical probability of spinning various colors. They then work toward determining the probability of having one land on one color and the other landing on another color, and discuss their strategy for this task.

Answers: $1/4$; $1/4$; $1/2$; $1/2$; $1/8$; using a tree diagram, writing out all the possibilities, multiplying 4×2

Station 3

Students will use a tree diagram to show the possible combinations when rolling two number cubes. They will use their diagram to answer questions about the likeliness of rolling different sums.

Answers: 11; No, because some sums repeat themselves; Rolling a 7: the sum is 7 six times.

Station 4

Students find the theoretical probability for drawing cubes out of a bag. They then consider how the probabilities would be affected if everything in the bag was doubled, and draw conclusions.

Answers: $1/5$; $1/5$; $2/5$; because there are twice as many yellow cubes as any other color; $2/10$ or $1/5$; $4/10$ or $2/5$; the probabilities do not change

Materials List/Set Up

Station 1	2 number cubes with the following sides: (4, 2, 2, 1, 3, 3) and (6, 1, 3, 5, 4, 8)
Station 2	none
Station 3	ruler
Station 4	bag containing one each of red, blue, and green cubes, and two yellow cubes

Discussion Guide

To support students in reflecting on the activities, and to gather formative information about student learning, use the following prompts to facilitate a class discussion to "debrief" the station activities.

Prompts/Questions

1. In what real-life situation might it be good to know the theoretical probability of an event happening?

2. You have 10 blocks in a bag—5 red, 2 yellow, 1 blue, and 2 green—and figured out the probability of drawing a red block. Now you triple the number of blocks in the bag (15 red, 6 yellow, 3 blue, and 6 green). How does this affect the theoretical probability?

3. Make a general statement about what happens to the probability of an event happening (pulling out a red block, for example) if we multiply everything each choice by the same number.

4. How can tree diagrams help you determine theoretical probabilities?

Think, Pair, Share

Have students jot down their own responses to questions, discuss their responses with a partner (who was not in their station group), and then discuss as a whole class.

Suggested Appropriate Responses

1. when buying a lottery ticket—you might decide not to since the odds are against you

2. It does not affect it.

3. The probability does not change.

4. They show you all the possible outcomes.

Possible Misunderstandings/Mistakes

- Not understanding that the ratio $^2/_5$ is equivalent to $^4/_{10}$
- Miscounting the number of answers that satisfy the probability you are looking for
- Flipping the ratios (e.g., $^5/_2$ instead of $^2/_5$)

Data Analysis and Probability
Set 4: Theoretical Probability

Station 1

At this station, you will find a very special pair of number cubes.

The table below shows the values on each number cube. Fill in the possible sums.

	1	2	2	3	3	4
1						
3						
4						
5						
6						
8						

What is the theoretical probability of rolling a 6? Write your solution as a ratio. _____

What is the theoretical probability of rolling an 8? Write your solution as a ratio. _____

Below is a table of all the sums possible when rolling two regular number cubes (sides 1–6).

	1	2	3	4	5	6
1	2	3	4	5	6	7
2	3	4	5	6	7	8
3	4	5	6	7	8	9
4	5	6	7	8	9	10
5	6	7	8	9	10	11
6	7	8	9	10	11	12

What is the theoretical probability of rolling a 6? Write your solution as a ratio. _____

What is the theoretical probability of rolling an 8? Write your solution as a ratio. _____

Look at both tables closely. What do you notice? _____

_____

_____

Data Analysis and Probability
Set 4: Theoretical Probability

Station 2

At this station, you will find two spinners. One spinner is evenly divided into four colors—red, green, blue, and yellow.

Work with your group to determine the following probabilities. Write them as a ratio.

P(yellow) = _____

P(red) = _____

The second spinner is evenly divided into two colors—black and white.

Work with your group to determine the following probabilities. Write them as a ratio.

P(black) = _____

P(white) = _____

What is the probability of spinning yellow and white? Use a tree diagram to draw out the different options if necessary.

What strategy did you use to determine the probability of spinning yellow and white?

_____

_____

Data Analysis and Probability
Set 4: Theoretical Probability

Station 3

At this station, you will use a tree diagram to help you figure out the theoretical probability of different totals you could get when rolling two number cubes.

Draw a tree diagram to show all the possible combinations of rolls if your number cubes are both numbers 1–6.

Write the probability of getting each total as a ratio on your tree diagram.

How many different possible totals can you get by rolling two number cubes? _____

Are all the totals equally likely? How do you know? _____

_____

_____

Which outcome is most likely? How do you know? _____

_____

Data Analysis and Probability
Set 4: Theoretical Probability

Station 4

At this station, you will be working on determining theoretical probability. You will see a bag with five cubes in it—one green, one blue, one red, and two yellow.

You should imagine what would happen if you pulled out one cube. Find the following theoretical probabilities. Be sure to write the probabilities as ratios.

P(green) = _____

P(blue) = _____

P(yellow) = _____

Why is the probability of pulling out a yellow different from the probability of drawing another color?

_____

_____

What would the probability of drawing a red be if we doubled the blocks that were in the bag (2 green, 2 blue, 2 red, and 2 yellow)? _____

What would the probability be of drawing a yellow in this same situation? _____

What do you notice about the probabilities if we double everything in the bag? _____

_____

_____

Data Analysis and Probability

Goal: To provide opportunities for students to develop concepts and skills related to understanding experimental probability and theoretical probability

NCTM Standards, Grades 6–8

Data Analysis and Probability

Understand and apply basic concepts of probability: use proportionality and a basic understanding of probability to make and test conjectures about the results of experiments and simulations.

Student Activities Overview and Answer Key

Station 1

Students will predict the number of times they will pull a red, green, and blue block from a bag. They will use theoretical probability to make this prediction. Students will then draw a cube from the bag and replace it 36 times in intervals of 12. They will compare their experimental results along the way with the theoretical results, and draw conclusions.

Answers: First table—4, 8, 12; second set of tables—answers vary depending on what blocks are pulled out; answers vary depending on what blocks are pulled out; answers may vary, but most likely students will be closer to their prediction for 36 turns than for 12 turns; Yes, the more turns we have, the closer the experimental probability is to the theoretical probability.

Station 2

Students will work with a list of randomly generated numbers. They predict what the mean should be based on the experimental probability of getting each number. They then compare their prediction with the actual average in a series of intervals. They will compare their experimental results with the theoretical result, and draw conclusions.

Answers: 5; table—answers will vary; The mean should approach the prediction as the amount of numbers increases; The experimental mean should be 5.

Station 3

Students will find the theoretical probability for rolling a 5 on a number cube. They will then each roll the number cube 20 times and compare the experimental probability to the theoretical probability. They will repeat this comparison after combining group data. Students then draw conclusions about sample size and experimental probability.

Answers: $1/6$; answers will vary; answers will vary; answers will vary—probably the second number because there was a larger sample size, and as sample size increases, experimental probability approached theoretical probability

Station 4

Students individually flip a coin 50 times to generate data. They then combine their data and look at the percentage of heads and tails that come up. Students discuss what the breakdown would look like if they flipped the coin 100,000 times.

Answers: Answers will vary; answers will vary; about 50% each because the experimental probability approaches the theoretical probability when sample sizes are very large

Materials List/Set Up

Station 1	3 blocks (red, green, and blue); a bag
Station 2	list of 100 randomly generated numbers (1–9)
Station 3	number cubes for each group member (1–6)
Station 4	coin for each group member

Discussion Guide

To support students in reflecting on the activities, and to gather formative information about student learning, use the following prompts to facilitate a class discussion to "debrief" the station activities.

Prompts/Questions

1. Why do we need a large sample size for the experimental probability to approach the theoretical probability?

2. Why do small samples differ from the theoretical probability?

3. Why is it important that we use random numbers or randomly pull a block out of the bag?

4. If you flipped a coin 1,000,000 times, how many times do you think heads would come up?

Think, Pair, Share

Have students jot down their own responses to questions, discuss their responses with a partner (who was not in their station group), and then discuss as a whole class.

Suggested Appropriate Responses

1. Large sample sizes account for any unusual outcomes that small sample sizes do not (e.g., rolling five 6s in a row).

2. Small samples can be affected by oddities in data.

3. Theoretical probability is not biased, so we should try to eliminate bias from experimental probability.

4. about 500,000 times

Possible Misunderstandings/Mistakes

- Difficulty dealing with a large amount of data

- Expecting small samples to look like the theoretical probability

- Thinking that when they have 30 or more pieces of data, the experimental probability will look exactly like the theoretical probability

---

NAME:

Data Analysis and Probability
Set 5: Experimental Probability

Station 1

At this station, you will find three blocks in a bag—one red, one green, and one blue. Imagine that you pull out a block, record what color it is, and put it back in.

Work with your group to determine how many times you would expect to pull out the green block if you went through this routine 12 times, 24 times, and 36 times. This should be the theoretical probability. Record your predictions in the table below.

Number of Times You Choose a Block	Prediction for the Number of Greens You Will Choose
12	
24	
36	

Now we are going to test our predictions. Place the three blocks in the bag. Without looking, take one block out. Record the color of the block you pulled out in the table below. Also, total the number of green blocks you pulled out after every 12 turns. Then answer the questions on the next page.

Turn	1	2	3	4	5	6	7	8	9	10	11	12	Total Green 1–12	Total Green 1–12
Color														

Turn	13	14	15	16	17	18	19	20	21	22	23	24	Total Green 13–24	Total Green 1–24
Color														

Turn	25	26	27	28	29	30	31	32	33	34	35	36	Total Green 25–36	Total Green 1–36
Color														

continued

145

© 2008 Walch Education   Station Activities for Mathematics, Grade 6

Data Analysis and Probability
Set 5: Experimental Probability

Station 1, *continued*

How close were your original predictions? _____
_____

Were you closer with your guess about how many green blocks would be pulled after 12 turns or after 36 turns?

_____

Do you think you would be even closer if you had to estimate how many green blocks you would pull out after 300 turns? Why or why not?

_____

_____

_____

Data Analysis and Probability
Set 5: Experimental Probability

Station 2

At this station, you have a list of 100 randomly generated numbers. Each number, 1–9, has an equally likely chance of coming up.

If the experimental probability of a number coming up is $\frac{1}{9}$ for every number, what do you think the mean of your list of numbers will be?

_____

Use the list of randomly generated numbers to fill in the table below. Under "Sum," write the total sum of the first 25, 50, 75, and 100 numbers. For "Mean," determine the mean of the set of data up to that point.

Amount of Numbers	Sum	Mean
25		
50		
75		
100		

What do you notice about your mean prediction as the amount of numbers you look at increases?

_____

_____

What do you think would happen if you found the mean of 10,000 randomly generated numbers (1–9)? Explain.

_____

_____

Data Analysis and Probability
Set 5: Experimental Probability

Station 3

At this station, you will find enough number cubes for each group member. Each group member should take one number cube. Work as a group to answer the following question.

What is the probability of rolling a 5? _____

Now each group member will roll a number cube 20 times on his or her own, and record the rolls in the table below.

	Tally	Total Number
1		
2		
3		
4		
5		
6		

What was the experimental probability of rolling a 5? _____

Combine data with your group.

	Total Number
1	
2	
3	
4	
5	
6	

What is the experimental probability of rolling a 5? _____

Which experimental probability was closer to the theoretical probability? Why?

_____

_____

Data Analysis and Probability
Set 5: Experimental Probability

Station 4

You will find a coin for every group member at this station.

Each group member should flip the coin 50 times, and record his or her data in the table below.

	Tally	Total Number
Heads		
Tails		

Now combine data with your group members.

	Total Number
Heads	
Tails	

What percentage of the flips was heads? _____

What percentage of the flips was tails? _____

What do you think the percentages would be if you flipped the coin 100,000 times? Why?

_____

_____

extending and enhancing learning

Let's stay in touch!

Thank you for purchasing these Walch Education materials. Now, we'd like to support you in your role as an educator. **Register now** and we'll provide you with updates on related publications, online resources, and more. You can register online at www.walch.com/newsletter, or fill out this form and fax or mail it to us.

Name _____ Date _____

School name _____

School address_____

City _____ State _____ Zip _____

Phone number (home) _____ (school) _____

E-mail _____

Grade level(s) taught _____ Subject area(s) _____

Where did you purchase this publication? _____

When do you primarily purchase supplemental materials? _____

What moneys were used to purchase this publication?

[　] School supplemental budget

[　] Federal/state funding

[　] Personal

[　] Please sign me up for Walch Education's free quarterly e-newsletter, *Education Connection.*

[　] Please notify me regarding free *Teachable Moments* downloads.

[　] Yes, you may use my comments in upcoming communications.

COMMENTS _____

_____

_____

_____

Please FAX this completed form to 888-991-5755, or mail it to:

Customer Service, Walch Education, 40 Walch Drive, Portland, ME 04103